QUANTUM
PHYSICS
IN MINUTES

GEMMA LAVENDER

QUANTUM PHYSICS IN MINUTES

GEMMA LAVENDER

Quercus

CONTENTS

Introduction   6
The birth of quantum physics   8
Energy levels and spectral lines   56
Particle physics   88
The wave function   142
The language of quantum physics   184
Quantum physics and the Universe   212
The theory of everything   252

Multiverses 286
The spooky Universe 304
Quantum applications 326
Quantum biology 356
Quantum computing 370
The future of quantum physics 390
Glossary 408
Index 412
Acknowledgements 416

Introduction

The world of quantum physics is an amazing place, where subatomic particles can do seemingly miraculous things. They can disappear from one location and spontaneously appear in another, or communicate with each other instantly despite being on opposite sides of the Universe. Real particles can mix with virtual ones that 'borrow' energy from the Universe, and these interactions govern the fundamental forces that bind atoms and molecules together, creating the structure of matter itself. To the uninitiated, it seems like magic.

Quantum physics pushes us to the boundary of what we know about physics, and scientists differ in their interpretations of what it all means. The one thing everyone seems to agree on is that on the smallest scales, nature is probabilistic – God really does play dice. For example, probability determines a particle's most likely location, or its most likely energy, momentum, or numerous other fundamental qualities. For some scientists, this is the extent of the meaning behind quantum physics. For others, it implies a multiverse of parallel realities where every possibility is played out. There's no direct evidence yet that this 'many-worlds' interpretation is correct, but the mathematics certainly suggests it is possible.

The true meaning of quantum physics may be still up for debate, but its myriad applications are far more concrete. All of the electronics in our computers, phones, televisions and tablets operate thanks to quantum principles. Lasers could not exist without the quantized energy levels in atoms. MRI medical scans utilise quantum mechanisms in action within your own body, and computers built around the principles of quantum physics might soon be solving problems much faster than any computer currently in existence. Quantum physics is also a step towards the ultimate theory of everything. It casts light on the origin of the Big Bang and the large-scale structure of the Universe, and some scientists controversially suggest that even human consciousness is quantum mechanical in nature.

Quantum physics is science, not magic. Yet what it can do is indeed magical and, by seeking to understand it, we find ourselves delving into the very fabric of nature and reality.

What is quantum physics?

Quantum physics describes the science of the very small, things tinier than billionths of a metre, on the scale of atoms, subatomic particles and the wavelength of light. It also shows how many properties are 'quantized' on these tiny scales, subdivided in discrete units rather than being continuously varying quantities. In our everyday world, it's hard to imagine the properties found in this microscopic world. For example, there are particles like electrons that have no physical dimensions, and others with no mass. Strangest of all, however, is the notion that particles can act like waves and waves can act like particles. This simple yet confounding fact lies at the heart of quantum physics and everything that subsequently flows from it.

It took scientists a long time to accept this bizarre idea, and the revolution that followed had a profound effect on modern science. Yet the discovery of quantum theory had its roots in a much older debate – the centuries-long argument over whether light is made from particles or waves.

'Anyone who is not shocked by quantum theory has not understood it.'

Attributed to Niels Bohr

Is light a wave?

Quantum theory has its roots in a fierce and long-running debate over the nature of light. The question as to whether light is made from particles or waves dominated science in the late 17th century. In 1678, Dutch scientist Christiaan Huygens popularized the hypothesis that light propagated in the form of a wave (based on earlier ideas by philosopher René Descartes).

Of course, waves (ranging from tidal waves in water to sound waves in air) need a medium through which to propagate. It was clear that light waves were not using air as a medium – space was known to be airless, and yet we can still see the light of the Sun, stars and planets. To get around this, Huygens hypothesized a medium that he called the 'luminiferous aether'. He neglected to explain exactly what this aether was, beyond it being weightless, invisible and apparently everywhere. Unsurprisingly, many scientists, key among them Isaac Newton, were unconvinced by Huygens' wave theory. Instead, they argued that light must be made from particles.

Studying the motion of water waves reveals aspects of wave behaviour, such as diffraction, that are also shared by light.

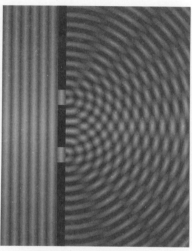

Diffraction after parallel waves pass through a narrow slit.

Interference pattern between two diffracted waves.

Is light a particle?

Influential English physicist Isaac Newton proposed a model
of light as discrete particles (so-called 'corpuscles'). It was
based not only on objections to the wave theory of Huygens,
but also upon observation. Newton pointed to the way in
which light is reflected from a mirror: waves do not travel in
the straight lines needed to create a reflection, but particles
do. Furthermore, Newton explained refraction (the bending
of light in certain materials, such as water) as the effect of
a medium attracting particles of light and speeding them up.
Finally, step outside on a sunny day and you will see that there
are sharp edges to your shadow, whereas if sunlight were
made of waves, your shadow would surely be fuzzy.

Newton's model became the leading theory of light, but it
wasn't well received by everybody, and his rival Robert Hooke
was one influential voice who still favoured the wave theory.
Then, in 1801, long after Newton's death, the double-slit
experiment seemed to disprove corpuscles once and for all.

Newton erroneously used the refraction of light passing through a prism to argue for its particle-like nature, but his discovery that white light can be split into many colours still led to important breakthroughs including the entire field of spectroscopy.

The double-slit experiment

Despite the success of Isaac Newton's corpuscular theory of light, the rival wave theory retained some proponents and, at the beginning of the 19th century, Englishman Thomas Young appeared to disprove Newton with an experiment that is replicated by high-school students to this day.

Young's experiment involves shining sunlight through a barrier containing two thin slits and onto a screen. Once through the slits, the light creates two spreading diffraction patterns, which begin to overlap and interfere with one another. Where a trough in one wave coincides with the peak of another, it causes them to cancel out, so that when the light finally reaches the screen, the cancelled waves leave dark bands known as 'interference fringes'. Since only waves can interfere in this fashion, Young concluded that light must be made from waves. By studying how the different colours within sunlight formed different fringe patterns, he was even able to estimate the wavelengths of the various colours.

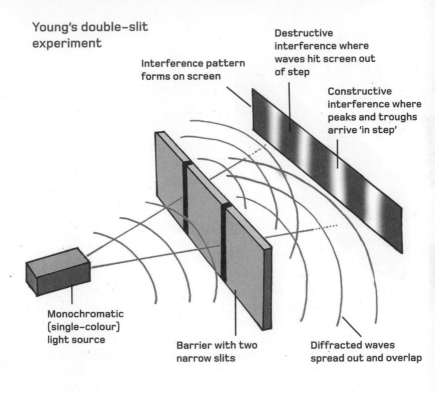

Young's double-slit experiment

Interference pattern forms on screen

Destructive interference where waves hit screen out of step

Constructive interference where peaks and troughs arrive 'in step'

Monochromatic (single-colour) light source

Barrier with two narrow slits

Diffracted waves spread out and overlap

The Michelson– Morley experiment

Thomas Young's proof that light is a wave implied that the light-carrying medium or aether proposed by Huygens (see page 10) must be real too, yet 19th-century scientists struggled to detect it. In 1887, American physicists Albert Michelson and Edward Morley set out to settle the question using an ingenious and highly sensitive experiment.

Theory held that the aether was stationary in space, so Earth's motion would result in the speed of light appearing faster in the direction of motion compared to a perpendicular direction. Michelson and Morley built a device called an interferometer to send beams of light from a single source along perpendicular paths before reflecting and recombining them. If the speed of light varied between the paths, then the returning waves would slip 'out of phase' with one another, creating a pattern of interference fringes that shifted over time. But try as they might, Michelson and Morley found the speed of light was the same in all directions. The aether did not exist, so how could light be a wave?

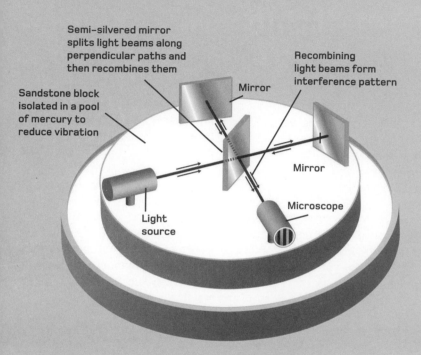

Semi-silvered mirror splits light beams along perpendicular paths and then recombines them

Recombining light beams form interference pattern

Sandstone block isolated in a pool of mercury to reduce vibration

Mirror

Mirror

Light source

Microscope

If the aether existed, then changes to the speed of light in different directions would cause interference patterns viewed in the microscope to shift over time.

Electromagnetism

If light really is a wave, then it seems reasonable to ask: *what exactly is doing the waving?* The properties of an apparently unrelated phenomenon, electromagnetism, ultimately proved to hold the answer.

In 1831, Michael Faraday discovered the phenomenon of induction, in which moving electrical currents 'induce' moving magnetic fields, and vice versa. Faraday's experiments (which still form the basis of electricity generation) showed a clear link between electricity and magnetism, but it was not until 1865 that James Clerk Maxwell set out a theoretical model for how induction and related effects took place. Maxwell's theory showed how oscillating, intertwined electric and magnetic fields can move through space as electromagnetic (em) waves. Crucially, he found that em waves moved freely through a vacuum, and propagate at a velocity of 300,000 kilometres per second (186,000 mps), exactly the same speed as light. If the aether didn't exist, then perhaps light was an electromagnetic wave?

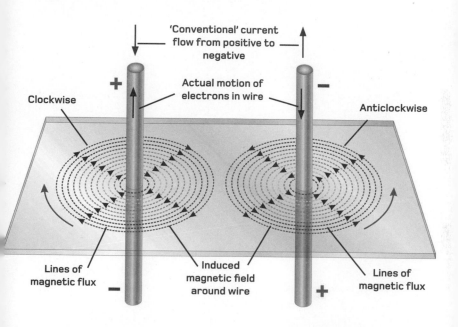

'Conventional' current flow from positive to negative

Actual motion of electrons in wire

Clockwise

Anticlockwise

Lines of magnetic flux

Induced magnetic field around wire

Lines of magnetic flux

The flow of electric current (a stream of negatively charged electron particles) through a wire induces a magnetic field around the conductor. Reversing the current reverses the direction of the magnetic field.

Maxwell's equations

To explain electromagnetism fully, James Clerk Maxwell appropriated a number of equations from other scientists and assembled them into a cohesive theory. His first equation describes how the strength of an electric field decreases with the square of distance. In other words, at twice the distance from the source, the field is four times weaker. The second equation describes the strength of magnetic fields and how they always follow closed loops between magnetic poles.

Maxwell's third equation describes how interactions between oscillating electric and magnetic fields can create 'electromotive force', which manifests as a voltage. Finally, the fourth equation describes how an oscillating electrical current can induce a magnetic field with a strength proportional to the size of the electrical current. Together, these equations describe how electromagnetic waves behave, while providing a theoretical explanation for exactly what a light wave is, how it propagates in a vacuum and how it interacts with electric and magnetic fields.

1. $\nabla \cdot \mathbf{E} = \dfrac{\rho}{\varepsilon_0}$

ε_0 = permittivity of free space

μ_0 = permeability of free space

2. $\nabla \cdot \mathbf{B} = 0$

3. $\nabla \times \mathbf{E} = -\dfrac{\partial \mathbf{B}}{\partial t}$

Rate of change of magnetic flux

Rate of change of electric flux

4. $\nabla \times \mathbf{B} = \mu_0 \left(\mathbf{J} + \varepsilon_0 \dfrac{\partial \mathbf{E}}{\partial t} \right)$

In Maxwell's equations, E represents the flux of the electric field, B the magnetic flux, ρ the charge within a volume of space, and J the current flowing in a conductor.

Thermodynamics and entropy

Alongside the discovery of electromagnetism, the study of energy in the form of heat led to another 19th-century scientific revolution. What became known as the laws of thermodynamics introduced several concepts that would prove critical to quantum theory.

The first law of thermodynamics explains how energy is conserved when heat is added to a closed system: the total energy of the system is equal to the heat supplied, less any work done (physical changes to the surroundings) as a result.

The second law, meanwhile, essentially describes how heat will always flow from hotter to colder systems. In fact, this law describes entropy, a measure of the amount of disorder in a system (illustrated opposite). The third law then explains how entropy approaches zero as the temperature within a system nears absolute zero. These notions of conservation of energy and entropy are discussed further on pages 82 and 320.

Increasing entropy

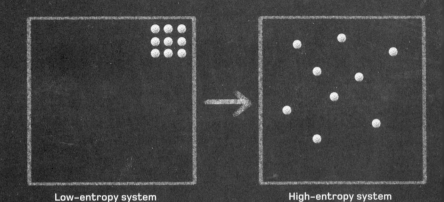

Low-entropy system   High-entropy system

The second law of thermodynamics states that systems will always tend to become less organized over time unless outside energy is applied. For example, a small splash of hot water will naturally tend to spread out through colder surroundings unless it is contained.

Black bodies

The study of the ways in which objects emit electromagnetic waves led, in the mid-19th century, to the idealized concept of a 'black body' radiator. This is an object that is both a perfect absorber and a perfect emitter of radiation. German physicist Max Planck found that the hotter the surface of a black body, the higher the energy of the light emitted. Hence, room-temperature objects glow mostly in infrared, while objects heated to thousands of degrees emit mostly visible wavelengths, and the hottest objects of all produce light in ultraviolet or even shorter wavelengths, such as X-rays.

A star is often considered the closest thing in nature to a perfect black body. Stars show the temperature—energy relationship in action: cooler stars emit more red light and infrared wavelengths, while hotter stars tend towards the blue and ultraviolet. Attempts to study what happens to black bodies at the highest temperatures were pivotal to the birth of quantum theory (see page 26).

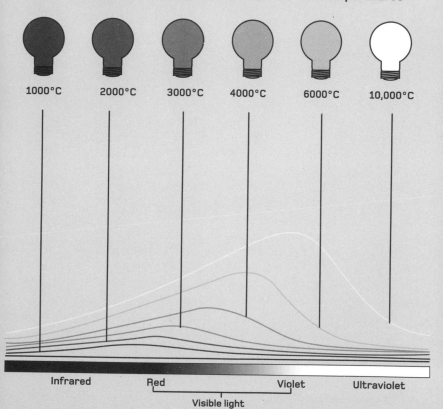

Light emission from black bodies of different temperatures

1000°C   2000°C   3000°C   4000°C   6000°C   10,000°C

Infrared   Red   Violet   Ultraviolet
                    Visible light

The ultraviolet catastrophe

In the late 19th century, physicists studying the behaviour of 'black bodies' at high temperatures found themselves faced with a problem: their models of the relationship between a black body's temperature and the distribution of radiation emitted from its surface fell apart at ultraviolet wavelengths. This was later nicknamed the 'ultraviolet catastrophe'.

Working to resolve the problem, around 1900 German physicist Max Planck found that two separate relationships described different parts of the energy distribution. An approximation derived by Wilhelm Wien in 1896 accurately described black-body radiation at high temperatures, while the Rayleigh–Jeans law (derived in 1900) showed that on the low-temperature end of the spectrum, the energy emitted by a black body is proportional to temperature divided by the wavelength to the power of four (as shown opposite). Planck now faced the challenge of reconciling these two apparently independent relationships.

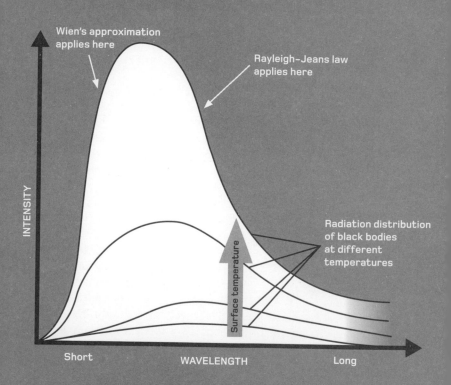

Quanta

Confronted with the problem of explaining black body radiation at high temperatures, Max Planck eventually discovered that he could explain the real-life distribution of radiation if he assumed that energy was not being released in continuous amounts, but was instead emitted in discrete bursts or packets of energy that he called quanta (*singular*: quantum).

Planck realized there was a relation between the energy and frequency of black body radiation, defined by the simple equation shown opposite. Here, E is the energy, f is the frequency and h is a constant of proportionality now known as Planck's constant (with a value of 6.626×10^{-34} joules per second). Planck assumed that the quantization of light was somehow a consequence of the way in which particles in a black body vibrate. It was not until 1905, however, that Albert Einstein adopted the idea of quantization, arguing that radiation was fundamentally divided into quantized packets called photons. Together, Planck's and Einstein's discoveries mark the birth of quantum physics.

$$E=hf$$

Discovery of electrons

At around the same time that scientists were homing in on the nature of light, the secrets of atomic structure were also beginning to unravel. The first hints of the existence of smaller particles inside atoms emerged from studies of a phenomenon known as cathode rays.

A cathode is a heated electrode that generates a beam of particles (in old television sets and laboratory displays, these were deflected using magnetic and electric fields to draw glowing images on a phosphorescent screen, as shown opposite). In 1897, English physicist J.J. Thomson determined that cathode rays were made of negatively charged particles with much smaller masses than atoms, being produced from inside them. The first subatomic particles ever to be discovered, these 'electrons' opened the way for an entirely new field of particle physics. At the time, scientists had little idea that their debates on the nature of light would soon collide with this new world of subatomic particles.

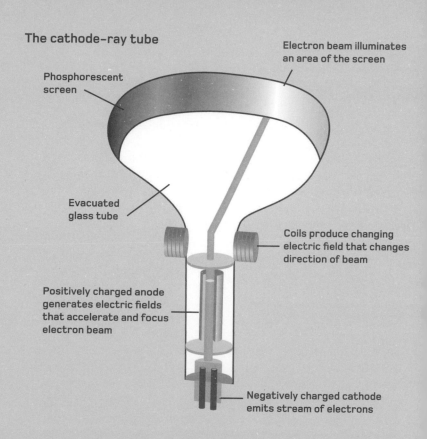

The cathode-ray tube

Phosphorescent screen

Electron beam illuminates an area of the screen

Evacuated glass tube

Coils produce changing electric field that changes direction of beam

Positively charged anode generates electric fields that accelerate and focus electron beam

Negatively charged cathode emits stream of electrons

The photoelectric effect

The photoelectric effect was crucial to both the conception and proof of Einstein's theory of photons (see page 34). Discovered by English engineer Willoughby Smith in 1873, the photoelectric effect involves the flow of electricity from some metals when they are illuminated under certain wavelengths of light. By the late 19th century, physicists knew enough to interpret this as the liberation of electrons from the surface of the illuminated metal, but the puzzling fact remained that, while high-frequency blue and ultraviolet light were efficient at knocking out electrons, even the most intense beams of red light could not cause electricity to flow.

Einstein realized that the photoelectric effect could be explained by interpreting light not as a continuous wave, but as discrete quantized packets similar to those used by Planck to escape the ultraviolet catastrophe (see page 26). Published in 1905, his theory predicted a relationship between the frequency of light and the energy of liberated electrons that was eventually proved in 1916 by American physicist Robert Millikan.

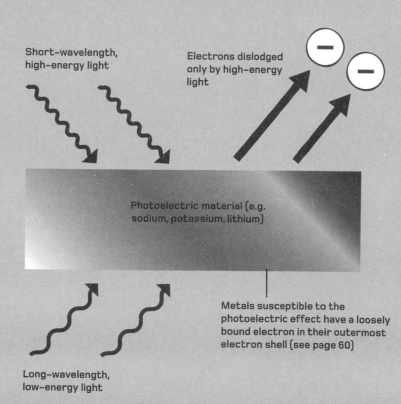

Short-wavelength, high-energy light

Electrons dislodged only by high-energy light

Photoelectric material (e.g. sodium, potassium, lithium)

Metals susceptible to the photoelectric effect have a loosely bound electron in their outermost electron shell (see page 60)

Long-wavelength, low-energy light

Einstein's photon theory

Albert Einstein's study of the photoelectric effect led him to some dramatic conclusions. Max Planck had shown that radiation from black bodies seemed to be released in small chunks whose energy content was related to frequency. Einstein now embraced the idea that this was an inherent aspect of light itself, rather than something entirely to do with the emission mechanism. According to him, light always came in quantized packets or photons, particle-like objects with energy proportional to their frequency.

This opened up a completely new approach to the photoelectric effect: atomic nuclei are surrounded by electrons in quantized energy levels, and it is these that interact with incoming photons. In order for an electron to escape from an atom, it must gain enough energy to leap the gap between energy levels. Einstein realized that individual photons either carry enough energy to bridge the gap, or they don't (unsuitable photons are deflected away). The deciding factor, then, is not the number of incoming photons (the intensity of the light) but their frequency (see page 28).

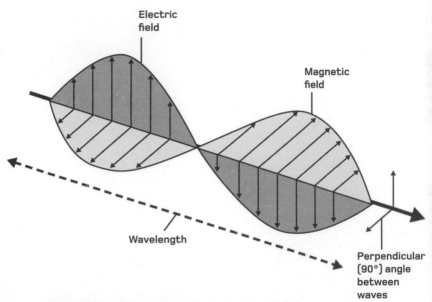

Electric field

Magnetic field

Wavelength

Perpendicular (90°) angle between waves

A single photon consists of a short burst of oscillating electric and magnetic fields moving through space and reinforcing each other by electromagnetic induction.

Compton scattering

In 1923, American physicist Arthur Compton demonstrated another effect that highlights the particle-like nature of electromagnetic radiation. Compton fired X-rays at carbon atoms and watched how individual photons rebounded or 'scattered' off electrons within them. X-ray photons have far more energy than is required to liberate an electron from an atom, so they have only to give up a little bit of energy to release an electron, retaining any remaining energy as they scatter away. Owing to this loss of energy, however, each photon now has a slightly lower frequency.

Compton related the process to billiard balls: one ball hits another, transferring some of its energy and momentum. Both balls recoil away, but the first ball moves more slowly than it did before the collision. This reflects the fact that momentum has to be conserved across an entire system during such collisions: if light behaves as though it has a momentum of its own, this adds to the evidence that it must be a particle, not a wave.

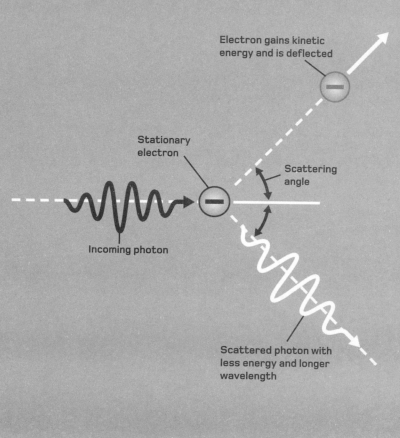

Electron gains kinetic energy and is deflected

Stationary electron

Scattering angle

Incoming photon

Scattered photon with less energy and longer wavelength

Wave–particle duality

By the early 1920s, the evidence that light had properties of both wave and particle was widely accepted, but the question of why light alone displayed this split personality remained a mystery. In 1924, French scientist Louis-Victor de Broglie suggested an explanation of sorts, namely that particles such as electrons *also* show dual aspects. He proposed that the 'wavelength' of a particle can be calculated by dividing the Planck constant (see page 28) by the particle's momentum (a property now known as the de Broglie wavelength, shown opposite).

Indeed, it turns out that *all* matter has an associated wavelength, and the shape of its wave (called the wave function) acts as a probability curve, with peaks in the wave at locations where the particle is most likely to be found. The higher the momentum, the shorter the de Broglie wavelength, so the wave aspect is only noticeable on atomic and subatomic scales. In contrast, the de Broglie wavelength of an Olympian running the 100-metre sprint is an undetectable 10^{-37} metres.

$$\lambda = \frac{h}{mv}$$

Electron diffraction

Direct proof of de Broglie's ideas about wave–particle duality was provided by Clinton Davisson and Lester Germer in 1929. Their experiment involved firing beams of electrons at a crystal of pure nickel. Because the de Broglie wavelength of electrons is much smaller than the wavelength of visible light, the narrow gaps between the crystal's atomic planes can act as a diffraction grating. Davisson and Germer measured interference fringes, similar to those created by light diffraction, in the intensity of electrons arriving on the other side of the crystal. The result was soon independently corroborated in a similar experiment by British scientist George Thomson.

The fact that electrons undergo diffraction not only showed conclusively that they have wavelike properties, but would also prove to have immense practical significance. The tiny wavelength of electrons allows us to use them to probe the structure of matter at much deeper levels than light microscopy.

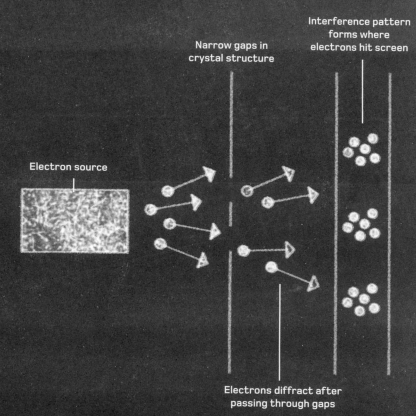

Rutherford's atomic structure

Wave–particle duality shows that quantum theory applies not only to light, but also to atoms and subatomic particles. So how did our understanding of these particles develop? Following his discovery of the electron (see page 30), J.J. Thomson proposed a simple model of the atom with negatively charged electrons embedded in a positively charged space, like plums in a pudding.

However, in 1908, Ernest Rutherford, Hans Geiger and Ernest Marsden discovered a more complex story. In a famous experiment, they fired radioactive alpha particles through a thinly beaten sheet of gold foil towards a phosphorescent screen that illuminated when struck by a particle. Most of the particles passed straight through the gold foil, but some had their paths deflected slightly, and others bounced straight back. Such behaviour was inexplicable in the 'plum-pudding' model, so Rutherford's team realized that most of the matter in an atom is compressed into a tiny central nucleus, now known to be composed of even smaller subatomic particles called protons and neutrons.

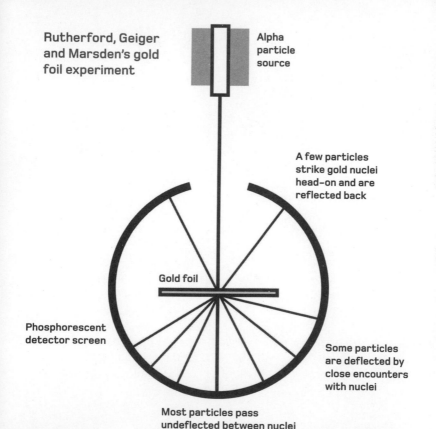

Rutherford, Geiger and Marsden's gold foil experiment

Alpha particle source

A few particles strike gold nuclei head-on and are reflected back

Gold foil

Phosphorescent detector screen

Some particles are deflected by close encounters with nuclei

Most particles pass undeflected between nuclei

Bohr's atomic structure

The atom suggested by Rutherford's gold-foil experiment is inherently unstable. In his model, electrons should lose energy, spiral in and collide with the positively charged nucleus, emitting light across a continuous range of wavelengths as they do so. Yet, in reality, atoms remain stable and light is emitted by electrons only in discrete quanta.

It was Danish physicist Niels Bohr who began to make sense of this by applying the nascent quantum theory to it. He depicted electrons as orbiting only in stable orbits, each with a specific energy level. For an electron to drop into a lower orbit, it must give up some energy, releasing a photon with an energy equal to the difference between the two orbits. Similarly, in order to jump to a higher orbit an electron must absorb a photon with sufficient energy. This is the basic theory behind the science of spectroscopy (see page 56), and the difference between energy levels is given by an equation called the Planck relation (shown on page 29).

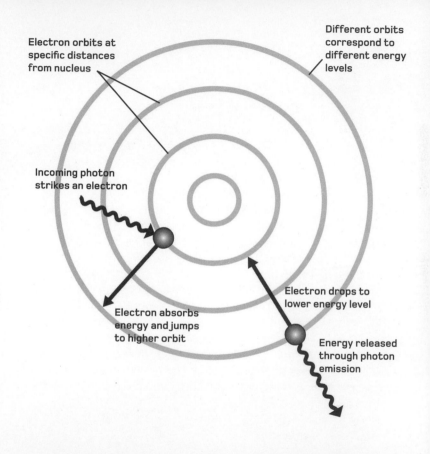

Electron orbits at specific distances from nucleus

Different orbits correspond to different energy levels

Incoming photon strikes an electron

Electron absorbs energy and jumps to higher orbit

Electron drops to lower energy level

Energy released through photon emission

The quantum mechanical atom

Despite the best efforts of Rutherford and Bohr, certain aspects of atomic structure remained a puzzle, and it wasn't until 1925 that the final pieces of the jigsaw slotted into place.

Wave–particle duality tells us that an electron can act as both a particle and a wave. If we imagine the orbits of electrons as concentric circles around the nucleus (analogous to the orbits of the planets around the Sun), then we should know where each electron is at any one time. However, if we then superimpose the probability wave of the de Broglie wavelength onto the electron orbits, we find that an electron could exist at any point along that wave: for want of a better word, its position becomes 'fuzzy'. German physicists Werner Heisenberg and Erwin Schrödinger, who first realized this, suggested that the electron orbits might better be considered as an electron 'cloud' or 'swarm'. Today, however, they tend to be referred to as 'orbitals'.

In the quantum model of the atom, electrons occupy diffuse orbitals rather than fixed orbits (see page 70 for an explanation of annotation)

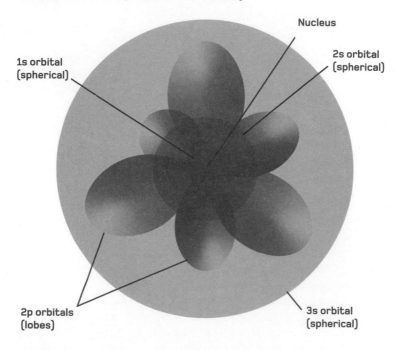

Nucleus

1s orbital
(spherical)

2s orbital
(spherical)

2p orbitals
(lobes)

3s orbital
(spherical)

Relativity

The early 20th century saw huge advances in physics. Alongside explorations of the subatomic and quantum realm, perhaps the biggest revolution of all came at the other end of the size scale, with Einstein's theories of special and general relativity (published in 1905 and 1915, respectively).

In fact, both theories would prove to have important implications for quantum physics. Development of the special theory (which explains the physics of objects moving at close to the speed of light but ignores situations of acceleration) led Einstein to the idea that mass and energy are equivalent, a keystone of quantum and particle physics (see page 50). The general theory, meanwhile, which considers accelerating 'reference frames', showed how gravity can distort space and make itself felt across vast cosmic distances. Einstein, however, wanted a more complete theory that could explain how gravity acts on very small scales, as well as at the very large, unifying relativity with the quantum world (see page 262).

Mass–energy equivalence

The most famous equation in all of physics tells us that mass and energy are equivalent. Einstein made this remarkable discovery while investigating how objects with mass (m) would behave when moving near the speed of light (c; the ultimate cosmic speed limit, reachable only by massless photons of light). The equation's huge scaling factor (the enormous speed of light multiplied by itself) reveals just how much energy is locked up in even a relatively small mass. For instance, a 1-kilogram (2.2-lb) bag of sugar locks up 3 million billion joules of energy. The more massive an object, the more energy it contains (and when an object moves, it carries even more energy – see page 82).

This mass–energy equivalence also holds for atoms and particles that operate under quantum rules. When physicists discuss particles, they don't talk about their mass in kilograms. Instead, they refer to their energies, measured in tiny units called electronvolts (eV). For example, the rest mass energy of an electron (the energy it has when it isn't moving) is 0.511 million eV.

$$E=mc^2$$

The Solvay Conference

The first steps towards unifying the disparate strands of quantum physics into a complete theory came at the 1927 Solvay Conference in Brussels, where 29 of the world's leading scientists assembled to wrestle with the subject. Among them were 17 Nobel Prize winners and giants in the world of quantum physics, including Niels Bohr, Arthur Compton, Marie Curie, Louis de Broglie, Paul Dirac, Albert Einstein, Werner Heisenberg, Wolfgang Pauli, Max Planck and Erwin Schrödinger.

There were (friendly) divides between these heroes and heroines of science. For example, Heisenberg considered the question of quantum physics settled, while Einstein was still groping for an explanation of why quantum mechanics worked at all. It was at this conference that he uttered his famous phrase, 'God does not play dice with the Universe,' in response to Heisenberg's uncertainty principle (see page 172). For Einstein, nothing should be left to chance.

Attendees at the 1927 Solvay Conference

The Copenhagen interpretation

Much of the theoretical groundwork laid for quantum physics in the 1920s was led by Niels Bohr (opposite) at the University of Copenhagen. Scientists including Paul Dirac, Erwin Schrödinger and Werner Heisenberg all came to Denmark to work with Bohr, and their collective efforts gave rise to the so-called 'Copenhagen interpretation'. This approach to quantum physics is a sort of ideology, claiming that everything we can definitely know about the behaviour of a quantum system emerges in the act of measuring it, and that without measurement, we are limited to describing a 'wave function' that predicts the probability of certain results.

Despite its popularity, however, this interpretation was not universally accepted. Other approaches have arisen since, from the many-worlds theory to the idea that only a conscious observer can cause a wave function to resolve into a single outcome (see pages 286 and 392). However, it was via breakthroughs made using the Copenhagen interpretation that scientists finally found the tools they needed to work within the bizarre world of quantum physics.

Spectroscopy

The science of spectroscopy is the analysis of the precise wavelengths at which materials and objects emit, absorb or reflect radiation. It is a hugely powerful tool, used today in a whole range of fields aside from astronomy, including medical research, materials science and chemical analysis. For example, although the element helium accounts for nearly a quarter of all atoms in the Universe, it was unknown until 1868, when astronomer Norman Lockyer identified a prominent 'gap' in the Sun's light output at a wavelength of 588 nanometres (billionths of a metre), and realized it was the signature of a new element in the Sun's atmosphere absorbing light.

Spectroscopy owes its power to the fact that the wavelengths of light emitted or absorbed by atoms are intimately linked to their internal structure, and are therefore dictated by quantum interactions happening between electrons at various energy levels. As such, it's an ideal proving ground for discovering and understanding many aspects of the quantum world.

Types of spectra

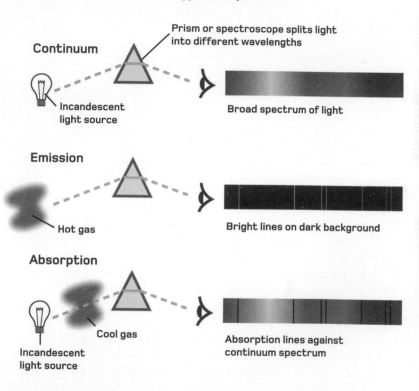

Continuum

Prism or spectroscope splits light into different wavelengths

Incandescent light source

Broad spectrum of light

Emission

Hot gas

Bright lines on dark background

Absorption

Incandescent light source

Cool gas

Absorption lines against continuum spectrum

Atomic structure

The model of an atom as depicted by Rutherford and Bohr (see pages 42 and 44) is fairly simple, with an atomic nucleus at the centre, surrounded by electrons in their orbits.

The nucleus is the heart of the atom. It contains most of the atom's mass, and is made from one or more 'nucleons', which are either protons or neutrons. Both of these particle types are themselves composed of three smaller particles called quarks (see page 92). A proton has a positive charge, while a neutron is electrically neutral. Because charge is balanced within an electrically neutral atom, the positive charge on a proton is cancelled out by the negative charge of an electron orbiting it.

The simplest element – hydrogen – usually has atoms consisting of a single proton orbited by a lone electron. Helium consists of two protons and usually two neutrons orbited by two electrons. At the other extreme, the heaviest known element has 118 protons, 118 electrons and 176 neutrons.

Simple atomic structures

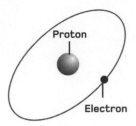

Hydrogen
1 proton, 1 electron

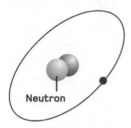

Deuterium (heavy hydrogen)
1 proton, 1 neutron, 1 electron

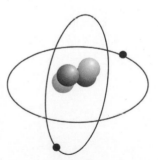

Helium
2 protons, 2 neutrons,
2 electrons

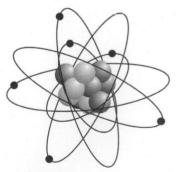

Carbon-12
6 protons, 6 neutrons,
6 electrons

Electron shells

Electrons flit around an atom's nucleus in shared 'shells'. The further away a shell is from the nucleus, the more electrons it can accommodate. The first shell is known as the 'K-shell' and it fits no more than two electrons in it. The second shell, called the 'L-shell', can include eight electrons; the third 'M-shell' up to 18 electrons; the fourth 'N-shell', 32 electrons; and so on. A handy formula allows you to calculate the total number of electrons an atom can contain in a given shell: $2(n^2)$, where n is the number of the shell, also known as the 'principal quantum number'. So in all the shells leading up to and including the M-shell, there can be $2 \times (1^2) + (2^2) + (3^2) = 2 \times (1 + 4 + 9) = 28$ electrons.

The greater its atomic mass, the more electrons an atom has, and hence more shells. In any atom, the outermost shell is known as the 'valence shell'. Since it directly interacts with other atoms, it is this shell that helps, above all, to define the chemical properties of that atom.

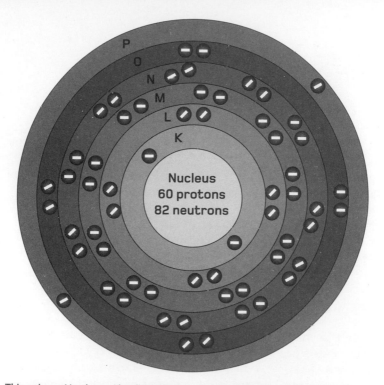

This schematic shows the distribution of electrons in an atom of neodymium (atomic number 60). The complexities of electron energy levels mean that its outermost 32 electrons are distributed through the N, O and P shells, rather than all collecting in the N shell.

Quantum numbers

The principal quantum number that describes the energy levels of electron shells is not the only way in which we define the quantum state – that is, the probability wave function – of an electron. Alongside it, there are other key quantum numbers that help define an electron's properties.

Electrons have a property called 'spin' (s; see page 102), which can have a quantum number of $\frac{1}{2}$ or $-\frac{1}{2}$. Meanwhile, the 'azimuthal' quantum number (l) describes the angular momentum of the electron (although an electron's mass is tiny at just 9×10^{-31} kg, it does exist, and therefore an orbiting electron must have angular momentum).

Finally, the 'magnetic' quantum number (m_l) describes the energy levels present in the 'subshells' of any given electron shell (see page 70). Within a magnetic field, m_l can also produce further shifts in electron energy, related to a phenomenon known as the Zeeman effect (see page 86).

Quantum numbers and atomic orbitals

Principal quantum number, n	Possible values of azimuthal quantum number, l	Subshell name	Possible values of magnetic quantum number, m_l	Number of orbitals in subshell *
1	0	1s	0	1
2	0	2s	0	1
	1	2p	1, 0, -1	3
3	0	3s	0	1
	1	3p	1, 0, -1	3
	2	3d	2, 1, 0, -1, -2	5
4	0	4s	0	1
	1	4p	1, 0, -1	3
	2	4d	2, 1, 0, -1, -2	5
	3	4f	3, 2, 1, 0, -1, -2, -3	7

* NB – Two electrons can coexist in each orbital if they have opposite spins.

Electron energy levels

Each electron shell within an atom exists at a different energy level, and the further from the nucleus an electron travels, the greater the energy it must have. Conversely, the closer it lives to the nucleus, the less energy it requires.

For example, consider an argon atom. This has 18 electrons, so entirely fills its K-, L- and M-shells. Electrons in the K-shell have -4,408 electronvolts (eV) of energy. The minus sign is explained by the fact that the potential energy of an electron only reaches zero at an infinite distance from the nucleus, so all the electron shells closer in are considered to have negative energy. The electrons in the L-shell have an energy of -1,102 eV and in the M-shell their energy is -489.78 eV, so the energy levels are getting higher (that is, closer to zero) with greater distance from the nucleus. For an electron to jump to a higher shell, it therefore must gain some energy by absorbing a photon. Conversely, to drop down from this excited state, it must lose some energy by emitting a photon.

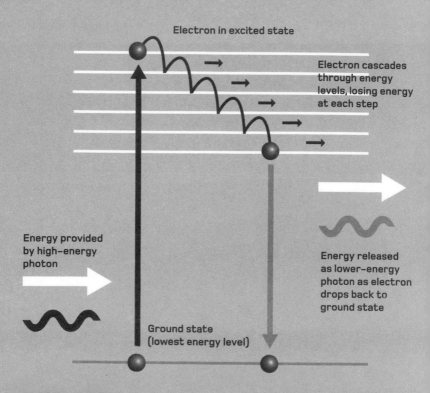

Electron in excited state

Electron cascades through energy levels, losing energy at each step

Energy provided by high-energy photon

Energy released as lower-energy photon as electron drops back to ground state

Ground state (lowest energy level)

Fluorescent materials glow when electrons excited by high-energy radiation drop back to their original state and re-emit photons of lower energy.

Calculating energy levels

When analysing spectra, physicists often need to calculate the approximate energy levels of electrons in a particular shell, and they can do this using the equation shown opposite. Here, E signifies the energy we are attempting to calculate, h is Planck's constant (see page 28), and c the speed of light, 2.998×10^8 m/s. R is a factor called the Rydberg constant with a numeric value of 1.097×10^7, Z is the atomic number (the number of protons or electrons in an atom) and n is the principle quantum number. In our previous example of the argon atom, $Z = 18$, so if we want to calculate the energy of the M-shell ($n = 3$), then we simply multiply everything together:

$$E = -6.626 \times 10^{-34} \times (2.998 \times 10^8) \times (1.097 \times 10^7) \times (18^2/3^2)$$

giving -7.845×10^{-21} joules. Quantum physicists, preferring to work in smaller units of energy, translate this to electronvolts (the energy needed to move a single electron across a 1 volt difference in electric potential), producing a result in this case of E= -489.78 electronvolts (eV).

$$E = -hcR\left(Z^2/n^2\right)$$

The ground state

The lowest-energy electrons in an atom (those in the innermost K-shell) are sometimes described as being in the ground state: their energy cannot get any lower and they have, in effect, hit the 'ground'. Those electrons in the higher electron shells with their extra energy are, on the other hand, described as being 'excited'.

The difference between ground-state and excited electrons is the key behind much of the quantum physics that happens in the electron shells. When an electron gains energy by absorbing a photon, it becomes excited and tries to jump up an energy level. Conversely, when an electron loses energy, it emits a photon and drops back down to a lower energy state. Why would an electron feel the need to drop down? All particles are at their happiest when they are as close to the ground state as they can get, so excited electrons are inherently unstable. If there is a vacancy in the electron shell immediately below it, an electron will shed excess energy to occupy that space.

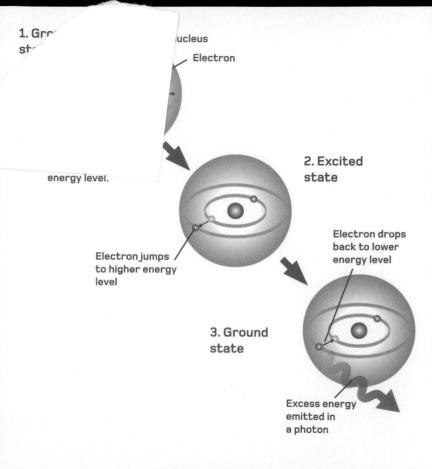

1. Gr... st...

Nucleus

Electron

energy level.

2. Excited state

Electron jumps to higher energy level

Electron drops back to lower energy level

3. Ground state

Excess energy emitted in a photon

Electron subshells

The equation given on page 67 might suggest that all electrons in the same shell have exactly the same energy, but this is not strictly true; as is often the case, quantum physics introduces some ambiguity. An electron shell is divided into a number of subshells, one of which will have the energy calculated using the equation, while the other subshells will have slight variations around that energy level. An electron in a given electron shell might exist in any of the subshells. The subshells are distinguished using more letters – s, p, d, f, and then following alphabetically, g, h, i, and so on. The innermost K-shell has just one subshell, known as 1s. The second, L-shell, has two subshells, 2s and 2p; the third, M-shell, has three subshells, 3s, 3p and 3d, and so on.

Subshells arise because the wave function of the electrons allows a little wiggle room in their spatial distribution – an electron will have a given probability of existing in one of the subshells. Which subshell in particular can be calculated using the famous Schrödinger wave equation (see page 156).

Generalized shapes of individual s, p, d and f subshells

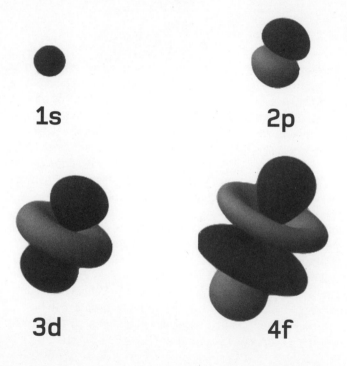

Quantum degeneracy

Mathematical models called 'harmonic oscillators' are ubiquitous in physics, and offer a way to describe many vibrating and wavelike phenomena. The simplest kind of oscillator is a mass attached to the end of a spring, bouncing up and down in one-dimension. However, consider a three-dimensional oscillation, one that's vibrating up and down, left and right, and back to front. If all three states of vibration oscillate with equal amplitude and energy, the system is described as 'degenerate'; fewer numbers are needed to describe it than we might expect.

Quantum physics has its own analogous concept of degeneracy that applies when more than one quantum state shares the same energy level in an electron shell. Quantum degeneracy describes how many quantum numbers can have the same energy, and is given by the square of the principle quantum number. In a hydrogen atom, a ground-state electron has a degeneracy of just 1, but if that electron is boosted into the L-shell, it's degeneracy becomes 4 (2^2); in the M-shell, 9 (3^2).

Energy levels in a single-electron (hydrogen) atom

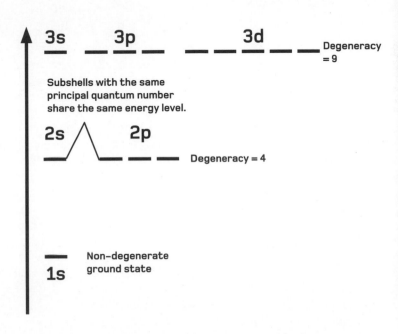

Hund's rules

In 1927, German physicist Friedrich Hund developed a set of rules that help to bring order to the potential chaos of electron configurations. He set out three rules to determine which configuration of an atom with several electrons has the lowest-energy ground state, and is therefore favoured.

The rules involve adding up the spin (s) and orbital angular momentum (l) of all individual electrons to find totals denoting S and L, respectively. Added together, these then give a 'total angular momentum quantum number' (J). The rules themselves are shown opposite, but it's their implications that matter most: Hund's rules imply that all electrons in singly occupied positions must be spinning in the same direction, and also that all empty positions must first be filled before electrons can begin pairing up.

This matters because the configuration of the electrons in the outermost shell defines an atom's chemical properties, so the order in which they occupy the positions is crucial in defining how one atom interacts with other atoms and molecules.

Hund's rules:

For a given electron configuration, the term with the lowest energy has the largest value of S.

For a given value of S, the term with the lowest energy has the largest value of L.

If the atom's outermost subshell is half-filled or less, then the lowest energy corresponds to the smallest value of J ($L+S$), but if it is more than half-filled, then the lowest energy corresponds to the *largest* value of J.

Pauli's exclusion principle

While Hund's rules describe the configuration of electrons, they don't offer an explanation as to why these configurations happen. That explanation lies in the famous exclusion principle discovered by prodigious Austrian physicist Wolfgang Pauli in 1925.

Pauli realized that the number of electrons in every fully occupied electron shell – two in the K-shell, six in the L-shell, ten in the M-shell, and so on – were identical to the number of possible different arrangements for the quantum numbers among electrons in that shell. He concluded that nature does not allow two electrons with identical quantum numbers to occupy the same shell; any that attempt to do so are excluded (forced into a different shell with a higher energy level). Thus, the structure of electron shells is prevented from collapsing. It later became clear that Pauli's exclusion principle applies not only to electrons, but also to neutrons and protons, with some fascinating consequences (see page 236).

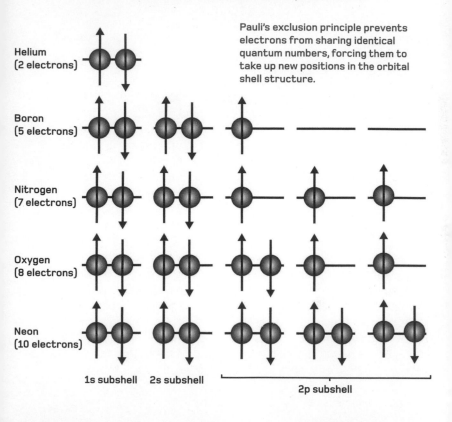

Electron distribution in some simple elements

Pauli's exclusion principle prevents electrons from sharing identical quantum numbers, forcing them to take up new positions in the orbital shell structure.

Helium
(2 electrons)

Boron
(5 electrons)

Nitrogen
(7 electrons)

Oxygen
(8 electrons)

Neon
(10 electrons)

1s subshell   2s subshell   2p subshell

Fraunhofer lines

When sunlight is split by a prism, we can see the visible region of the electromagnetic spectrum dispersed according to the different wavelengths in its light. In the early 19th century, scientists began noticing that this rainbow-like pattern of colours was crossed by dark lines. German optician Joseph von Fraunhofer studied these in depth, identifying over 500 lines, although thousands of these are now known.

Fraunhofer lines are produced by atoms in the Sun's atmosphere absorbing light emitted at its visible surface, and are a consequence of the discrete energy levels within atoms. To jump between levels, an electron must absorb a photon and take a precise amount of energy corresponding to a specific wavelength in the spectrum. When many atoms absorb similar photons they remove that wavelength from the broad 'continuum' of sunlight, creating dark absorption lines. Each line matches a specific transition in a particular atom, so scientists can use them to identify the atomic make-up of stars and other objects.

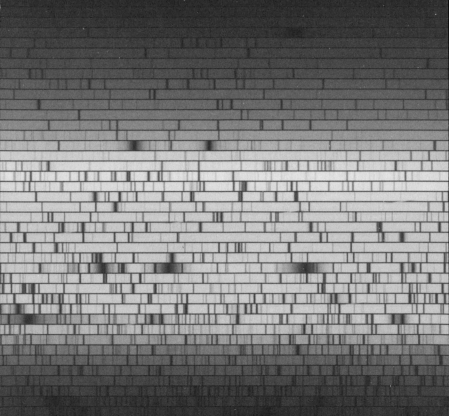

Emission lines

As well as dark absorption lines, electrons can also create bright 'emission lines' by *releasing* photons as they transition from an excited state to a less excited one. Emission lines have wavelengths equivalent to the difference in energy involved in a transition, and are produced by energized gases, for instance in neon strip lights or the nebulae surrounding newborn stars.

Hydrogen, the simplest and most common element in the Universe, has its own set of electron transitions giving rise to specific wavelengths of light. The Lyman series, discovered by Theodore Lyman in 1906, is produced by electrons dropping from various energy levels straight to the ground state. The transition from the L-shell to the K-shell, known as the Lyman alpha, corresponds to ultraviolet radiation at a wavelength of 121 nanometres. The transition from the M-shell to the K-shell is termed Lyman-beta, and so on. Meanwhile, the Balmer series, which includes both visible and ultraviolet lines, describes transitions from higher excited states down to the L-shell.

The Orion Nebula is an emission nebula—a cloud of hydrogen-rich gas excited by radiation from newborn stars.

Conservation of energy and momentum

The first law of thermodynamics (see page 22) is based on the idea that the total energy of a system is conserved; energy cannot be created or destroyed, but it can change from one form into another. For example, when an ice cube is heated and sublimates, an input of heat energy is transformed into kinetic (motion) energy in the molecules of water vapour.

Energy is not the only property in the Universe that is conserved; momentum is too. Think of balls on a pool table. When a player strikes the white ball with the cue, it gives the ball a certain momentum. When the white ball hits a stationary red ball with zero momentum, some of the white ball's momentum is passed to the red ball, while the white ball retains the rest of its momentum as it rebounds away. The total momentum of the white ball and the red ball after the collision is the same as the total momentum between the two before the collision. Although ice cubes and pool tables are everyday examples, the principles behind these conservation laws also hold in particle physics.

Elastic and inelastic collisions

Elastic collision preserves both momentum and kinetic energy

Mass m

Velocity = v

Mass m

m

Velocity = 0

m

Velocity = 0

m

Velocity = v

Total momentum = mv
Total kinetic energy = ½ mv²

Total momentum = mv
Total kinetic energy = ½ mv²

inelastic collision preserves momentum but not kinetic energy

Mass m

m

Velocity = v

Mass m

m

Velocity = 0

Particles stick together

Velocity = v/2

Total momentum = mv
Total kinetic energy = ¼ mv²
(Remaining energy transferred to other forms)

Forbidden transitions

Although nature as depicted by quantum physics is inherently uncertain and messy, it can occasionally be fussy about a few things – notably the conservation of properties such as energy, momentum (see page 82) and quantum states.

However, an electron doesn't inherently know what the rules are. When it absorbs a photon, it will naturally attempt to make a quantum jump to a higher energy state, even if that jump takes it into a quantum state that's already occupied or one that doesn't conserve quantum numbers. When an electron makes one of these 'forbidden transitions', it is swiftly forced to return to its original state. However, by then it is in some ways too late: the electron has already absorbed a photon, so we see a dark absorption line. When returning to its original state the electron also has to spit out a photon carrying the excess energy, so creating an emission line. Hence, absorption and emission lines can appear even when the laws of physics say that certain transitions shouldn't officially take place.

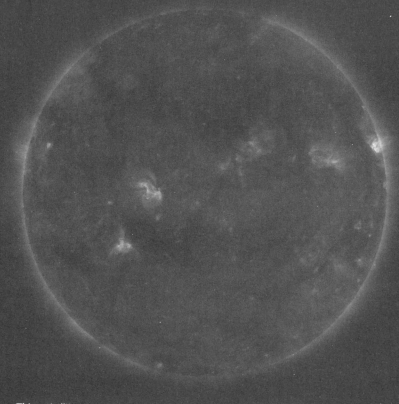

This satellite view of the Sun shows only light emitted by a forbidden transition of iron atoms.

The Zeeman effect

The concept of electron spin arises frequently in quantum physics (for example, in our discussions of quantum numbers and Hund's rules, see pages 62 and 74). As effective 'point masses' with negligible physical dimensions, electrons don't *actually* spin, but they still sport quantum properties that behave as though they were spinning, and this spin can be distinguished as either up or down (in terms of quantum numbers, spins of either ½ or -½).

Because electrons have negative electrical charges, their passage through an electric field generates a magnetic field, and the spin of the electron decides the field's magnetic polarity. Therefore, an electric field will deflect electrons with spin up in one direction, and those with spin down in a different direction. There is also a small but measurable energy difference between the two spins, which can result in the fine splitting of energy levels while under the influence of an electric field. This effect was first noted by, and subsequently named after, Dutch physicist Pieter Zeeman.

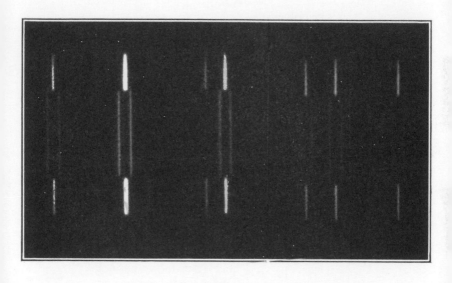

A beam of electrons fired through an electric field will produce two spots, not one.

The particle zoo

The protons, neutrons and electrons found inside everyday atoms are just the tip of the iceberg when it comes to the particle family tree. The 20th century saw an explosion of discoveries as particle accelerators ramped up in power and theoretical physics became more advanced. Research continues apace in the 21st century, significantly with the discovery of the famous Higgs Boson in 2013.

We describe our best picture of particle physics as the 'Standard Model'. Like a grand soap opera, the Standard Model's family tree is actually the story of three families – the fermions, the quarks and the so-called force carriers or bosons – and how they get along with one another. Relations between these particles are determined via fundamental forces known as the strong, weak and electromagnetic interactions (the fourth fundamental force, gravity, is negligible on particle scales). On these subatomic levels, quantum physics is king, moulding and influencing the particles to do its bidding.

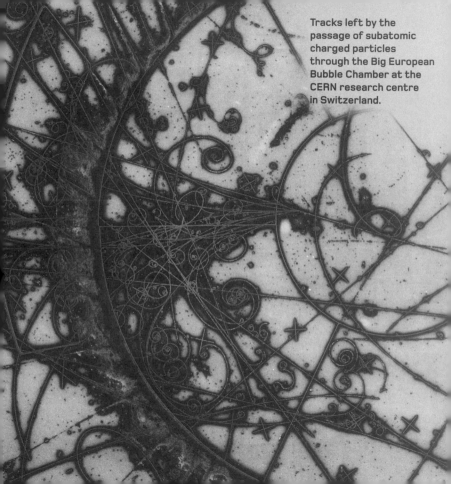

Tracks left by the passage of subatomic charged particles through the Big European Bubble Chamber at the CERN research centre in Switzerland.

The Standard Model

Physicists converged on the so-called Standard Model of particle physics in the 1970s, following several decades of research and discovery. The model describes particles at their most fundamental level – those that are indivisible or elementary, ground zero for the construction of matter (for example, quarks, which cluster together to form other particles, such as protons and neutrons).

However, the Standard Model is more than just an exercise in cataloguing particles; it's a model of how the particle world operates, describing how those particles interact in ways that ultimately enable them to create the world we see around us. We'll learn more about them and their interactions in the following pages. Yet there will also be unanswered questions, puzzles that the Standard Model cannot yet explain, such as the origin of the fundamental forces, how neutrinos are able to change from one type into another, as well as the nature of mysterious 'dark matter'.

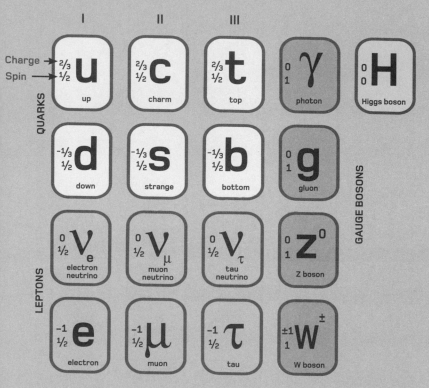

Quarks

The protons and neutrons found in each atomic nucleus are themselves made of tiny particles known as quarks. Murray Gell-Mann, the physicist most associated with their discovery, named them after a line in James Joyce's novel, *Finnegans Wake*.

Quarks are tiny, no more than a million-trillionths of a metre in size, each containing about one-third the mass of a proton or neutron. They are also unique in the Standard Model in that their charges are fractions of the charge on an electron, rather than whole-number multiples. They come in six flavours, somewhat whimsically referred to as up, down, strange, charm, top and bottom. The last three forms are highly unstable, and don't seem to play a role in the structure of matter.

Quarks bind together in clusters of two or three through a fundamental interaction known as the strong force (clusters of three form protons and neutrons). So powerful is this force that no quark has ever been observed in isolation.

The three generations of quarks

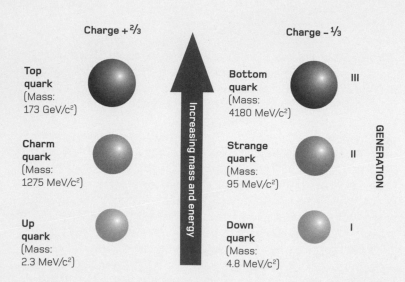

Hadrons

Particles built from quarks are known as hadrons. They can be further split into two families, namely baryons (the protons and neutrons of the atomic nucleus) with three quarks, and a variety of short-lived meson particles with two. The physics underlying their bonding is known as quantum chromodynamics, or QCD (see page 260). 'Chromo-' refers to colour (not *literal* colour, but a unique quantum property belonging to quarks).

Pauli's exclusion principle states that particles with identical quantum numbers cannot occupy the same space. The addition of a colour property allows quarks that would otherwise have the same quantum numbers to get around this. There are three 'colours': red, green and blue (plus antiquarks that are antired, antigreen or antiblue). Colours and their anticolours attract, and can bind two quarks into a meson. The three colours also attract each other, leading to baryons formed of one red, one blue and one green quark. A 'boson' particle called a gluon, meanwhile, transmits the strong force between quarks.

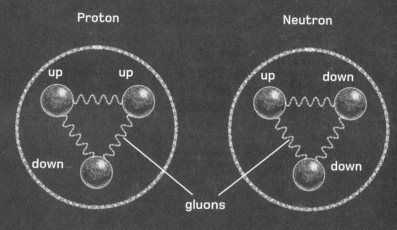

Proton

up   up

down

Neutron

up   down

down

gluons

2 up quarks,
1 down quark
Net charge +1
Total spin + ½

2 down quarks,
1 up quark
Net charge 0
Total spin + ½

Leptons

Electrons and the other members of their particle family are known as leptons. These are a different breed of particle to the hadrons, not least because they are indivisible: there are no quarks or other tiny particles making up leptons, so we say they are 'elementary'. Alongside the electron, which is vital to quantum physics, there are two other types of lepton: the tau particle and the muon. These are highly unstable and don't really play much of a role in normal particle physics.

Another form of lepton — the neutrino — holds the crown as the most bizarre particle that we know. Neutrinos fill the Universe, with trillions streaming through your body at this very moment. Yet these particles have barely any mass at all, and are able to oscillate between three 'flavours' named after their fellow leptons: electron, tau and muon neutrinos. Because they interact so weakly with other particles and have no electrical charge, neutrinos can only be detected indirectly in experiments built deep underground to shield them from interference.

The three generations of leptons

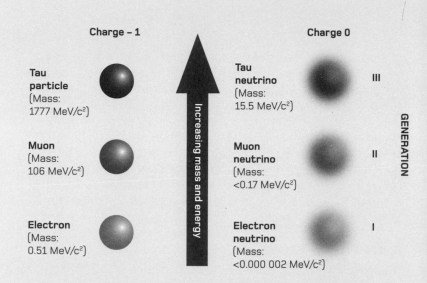

Charge – 1

Tau particle
(Mass: 1777 MeV/c²)

Muon
(Mass: 106 MeV/c²)

Electron
(Mass: 0.51 MeV/c²)

Increasing mass and energy

Charge 0

Tau neutrino
(Mass: 15.5 MeV/c²)

Muon neutrino
(Mass: <0.17 MeV/c²)

Electron neutrino
(Mass: <0.000 002 MeV/c²)

III

II

I

GENERATION

Dark matter

Despite the great strides made by physics in the past century, it's a striking fact that all types of visible matter account for just 4.9 per cent of the mass and energy in the Universe. The remainder is completely unknown, and the bulk of that is accounted for by an energy field known only as dark energy. Some 26.8 per cent, however, is composed of a substance called dark matter. This is referred to as dark because we cannot see it – it does not emit or absorb radiation of any sort, and we can only detect its gravity through the way it affects the motions of stars on the edges of galaxies, and of galaxies on the edges of galaxy clusters.

Dark matter is assumed to consist of subatomic particles of some description, that may be 'hot' or 'cold' (depending on how much energy they have). Neutrinos were once a popular candidate for hot dark matter, but dark matter is now mostly thought to be dominated by cold 'WIMPs' (weakly interacting massive particles), whose identity remains mysterious.

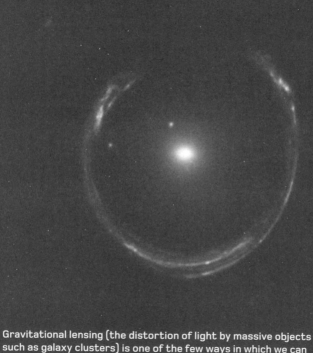

Gravitational lensing (the distortion of light by massive objects such as galaxy clusters) is one of the few ways in which we can directly observe the effects of dark matter.

Electric charge

The electric charge on a particle is, fundamentally, an indication of its susceptibility to electric and magnetic fields, and its ability to take part in electromagnetic interactions. Like energy and momentum, charge is conserved, meaning the total charge in a system is the same before and after an interaction between particles. In fact, charge can play a key role in such interactions, since it is susceptible to the electromagnetic force. Charge is also quantized: it comes in discrete 'packets' that are usually integer multiples of the charge of the electron or, in the case of quarks, simple fractions of the electron charge. Particle charge is never in random amounts, but increases in quantum leaps, so the charge that a particle has can be predicted by the Standard Model.

Particles such as electrons have a negative charge, while others like protons and positrons ('antielectrons') have a positive one. Neutrinos and neutrons, meanwhile, are electrically neutral, due either to an inherent lack of charge, or because the charges of their constituent particles cancel out.

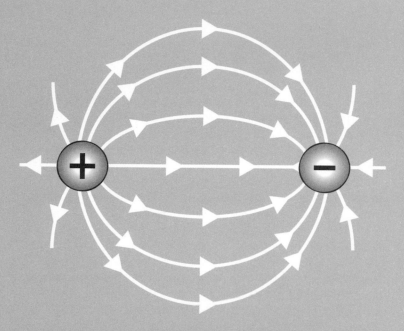

Every charged particle is surrounded by its own electric field, conventionally depicted as emerging from positive charges and flowing into negative ones. As with magnetic poles, like charges repel while unlike ones attract each other.

Particle angular momentum

Angular momentum is simply the momentum of something spinning on its axis or moving around something, as distinct from the momentum of its straight-line motion through space. Particles can have a real angular momentum L (a property that is both quantized and conserved during particle interactions – see page 82), but the calculations that describe a particle's angular momentum also describe another particle property known as 'spin'. This is important because spin is a quantum number, one of the key parameters describing the state of a particle.

Spin is a strange property; despite the name, it does not refer to actual physical rotation, yet it endows particles with a magnetic moment in the same way as if it did (see page 106). Calculating L involves multiplying a 'position operator' r by a 'momentum operator' p (operators are tools used to deal with vectors – see page 196). To calculate the *total* angular momentum of a particle (denoted J) we also have to add the total spin S, so total angular momentum $J = rp + S$.

Spin − ½

Spin + ½

Although spin is not a true
rotation like that of, for
example, a spinning top, it
is still useful to think of
particles as spinning in one
direction or another at
particular quantized rates.

Chirality and parity

Rotational symmetry plays an important role in particle physics. Systems that are identical after a 360-degree rotation of their wave function are said to have parity, whereas those that change are said to be chiral. Picture an electron with spin ½. If you rotate its wave function by 360 degrees, the result is not identical to the original, and spin changes to -½. These changes cause it to behave differently in terms of both quantum properties and interactions with other particles. An analogy is a mirror image: if an object is not identical to its mirror image, it too is chiral. Most letters of the alphabet, for instance, are chiral but a few, such as 'A' and 'H' have parity.

Chirality can be either left- or right-handed (referring to how the wave function is rotated), and 'handedness' has some important consequences. For example, only left-handed fermions (i.e. electrons) or right-handed antifermions (i.e. positrons) are capable of interacting with the weak force responsible for radioactive decay.

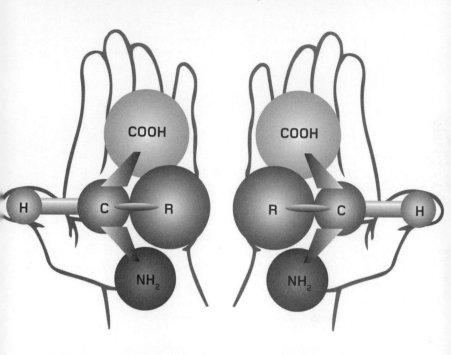

Just as the amino acids illustrated here are chiral, with distinct left-and right-handed forms, so subatomic particles can also be chiral. In both cases, this can have an important influence on their behaviour.

Magnetic moments

Faraday's laws (see page 18) tell us that a rotating electric field can induce a magnetic field. Some particles – electrons, for example – possess charge. They are essentially tiny packets of electrical current (the electricity we plug into in our homes is just a stream of electrons). Electrons also have inherent quantum spin, so their rotating electric fields should also generate a tiny magnetic field. Physicists refer to this field as an electron's magnetic dipole moment. The 'magnetic moment' of a regular bar magnet is calculated as the strength of the poles multiplied by the distance between them, so it depends in part on how large the bar magnet is. Calculating the magnetic moment of an electron is somewhat more complicated, but the magnetic moment is always 'antiparallel' to the spin angular momentum.

The term 'dipole' indicates that a field has both north and south poles. Magnetic monopoles with just one pole have not yet been found to exist in the Universe, though they are proposed by some advanced theories, such as superstrings (see page 274).

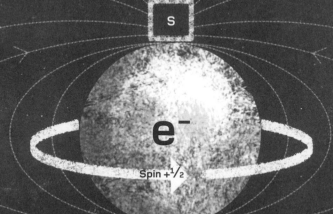

S

e⁻

Spin +¹/₂

N

The quantum spin of an electron generates
a rotating electric field which in turn gives
rise to a magnetic dipole field, similar to that
around a bar magnet.

Spin—orbit interactions

When an electron is locked in orbit around an atomic nucleus, the magnetic moment created by its quantum spin (see page 102) will interact with the magnetic field induced by its orbital motion around the nucleus. This spin—orbit interaction leads to the creation of unexpected features that can be detected within the spectral lines of an atom. The spin of the electron relative to the induced magnetic field can have two different orientations, either parallel to the direction of motion, or 'antiparallel'. Parallel spin has a marginally lower energy than would otherwise be expected, while antiparallel spin has slightly higher energy. This results in the atomic energy levels splitting in two, a phenomenon that physicists refer to as 'fine structure'.

There is also an opposite effect, where the nucleus itself has spin and, therefore, a magnetic moment, which interacts with the magnetic field induced by the electron. This results in 'hyperfine structure' – extremely small shifts in spectral lines that reveal secrets of the energy levels within the atom.

Fine structure features are responsible for the paler lines in this spectrum of deuterium.

Fermions

Particle physicists catalogue subatomic particles in many different ways. Just as particles can be associated with different 'families', such as baryons, leptons and mesons, so groups of families can be aligned under common characteristics, such as quantum spin. This gives rise to the 'fermions', which are particles with half-integer spin (in other words, spins of $\frac{1}{2}$, $\frac{3}{2}$, $-\frac{1}{2}$ or $-\frac{3}{2}$). The fermions include quarks, and hence encompass the baryons (see page 94) since the three quarks bound together in these particles result in a spin of either $\frac{1}{2}$ or $\frac{3}{2}$. Leptons, which have a spin of $\frac{1}{2}$, are also fermions, as are composites of particles made from an odd number of quarks, such as some atomic nuclei.

Fermions are named after Italian nuclear physicist Enrico Fermi. Their behaviour is described by a mathematical model called Fermi-Dirac statistics, and their most important property is that they adhere, without fail, to Pauli's exclusion principle, with important consequences for atomic structure (see page 76).

LEPTONS (Spin ½)			
Symbol	Flavour	Mass (MeV/c^2)	Charge
ν_e	Electron neutrino	< 0.000 002	0
e	Electron	0.51	-1
ν_μ	Muon neutrino	< 0.17	0
μ	Muon	106	-1
ν_τ	Tau neutrino	15.5	0
τ	Tau	1777	-1

QUARKS (Spin ½)			
Symbol	Flavour	Mass (MeV/c^2)	Charge
u	Up quark	2.3	$+2/3$
d	Down quark	4.8	$-1/3$
c	Charm quark	1275	$+2/3$
s	Strange quark	95	$-1/3$
t	Top quark	173,000	$+2/3$
b	Bottom quark	4180	$-1/3$

Bosons

Named after Indian physicist Satyendra Nath Bose, and described by a model called Bose-Einstein statistics, bosons are particles that have integer or whole-number spins (0, 1, 2, and so on.). For reasons not entirely understood, bosons happily flout Pauli's exclusion principle, so that many bosons can coexist in the same quantum state. This is important because it allows them to permeate space as so-called 'gauge bosons' – particles that transmit the four fundamental forces of nature that hold matter together and govern how particles interact (electromagnetism, the strong force, the weak force and gravity).

Only elementary bosons are force carriers: these include photons (carrying the electromagnetic force), gluons (the strong force), W and Z bosons (the weak force) and the hypothetical graviton (gravity). Other bosons include mesons (see page 94), whose two quarks have a combined spin of 0 or 1, and atomic nuclei such as helium-4 (which contains two protons and two neutrons, amounting to 12 quarks in all and hence a total spin of 6).

ELECTROWEAK FORCE				
Symbol	Name	Mass (GeV/c^2)	Charge	Spin
γ	Photon	0	0	1
W^+	W^+ boson	80.39	+1	1
W^-	W^- boson	80.39	-1	1
Z^0	Z^0 boson	91.188	0	1

STRONG FORCE				
Symbol	Name	Mass (GeV/c^2)	Charge	Spin
g	Gluon	0	0	1

HIGGS FIELD				
Symbol	Name	Mass (GeV/c^2)	Charge	Spin
H	Higgs boson	126	0	0

Bose–Einstein condensates

The fact that bosons don't obey Pauli's exclusion principle means there is no limit as to how many can be packed into the same energy level with the same quantum numbers. In the 1920s, Albert Einstein and Satyendra Nath Bose realized this could have some strange consequences. Their Bose–Einstein statistics described all the quantum states in which a gas of bosons could exist. Einstein wondered what would happen if those bosons were chilled to just a few degrees above absolute zero. He proposed that all the bosons would sink to the lowest possible energy level, creating a new form of matter called a Bose–Einstein condensate.

Condensates were finally produced in laboratories in the 1990s, and freakishly display quantized properties on a visible scale. For example, when helium-4 (a bosonic gas) is cooled to 2°C (3.6°F) above absolute zero, it starts to act as a superfluid – a liquid with no resistance to movement. Bose display many curious properties – they can even slow the passage of light to a crawl and even stop it, while when stirred, they form vortices that continue to swirl indefinitely.

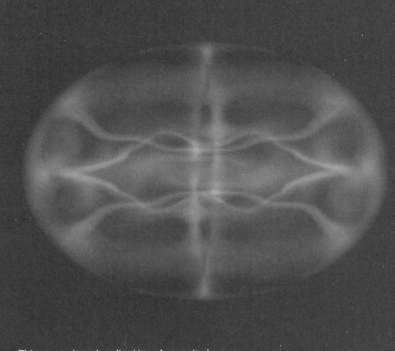

This computer visualization shows ripples travelling through a Bose–Einstein condensate at temperatures close to absolute zero.

The Large Hadron Collider

In order to detect new subatomic particles, physicists have built immense machines that smash particles together at ultrahigh speeds, creating higher and higher energies that briefly give rise to short-lived particles not usually found in nature. The biggest of these particle detectors is the Large Hadron Collider (LHC) at CERN, on the Swiss-French border near Geneva.

The LHC's ring-shaped underground tunnel, some 27 kilometres (16.8 miles) in circumference, is lined with 1,625 superconducting magnets that bend, accelerate and focus beams of hadrons ranging from isolated protons to heavier atomic nuclei and electrically charged ions. Two beams at a time race around in opposite directions, each loaded with up to 120 billion hadrons. Accelerating through the underground tubes, their speeds approach the speed of light before they smash into each other at energies of up to 13 trillion eV. Each year several hundred trillion collisions take place, recorded by seven separate experiments located in enormous chambers around the ring.

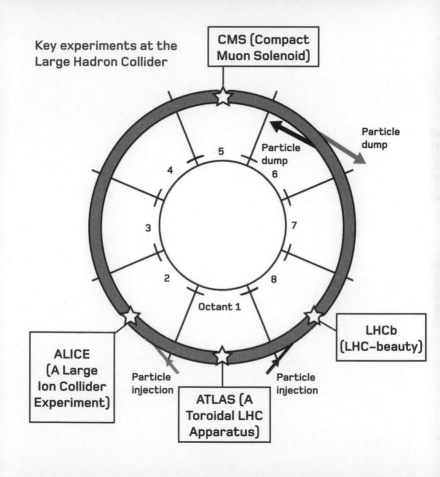

Key experiments at the Large Hadron Collider

CMS (Compact Muon Solenoid)

Particle dump

Particle dump

5

4

6

3

7

2

8

Octant 1

ALICE (A Large Ion Collider Experiment)

Particle injection

ATLAS (A Toroidal LHC Apparatus)

Particle injection

LHCb (LHC-beauty)

LHC discoveries

The Large Hadron Collider is designed to test all kinds of exotic particle physics theories, from ideas such as supersymmetry (see page 274) to extensions of the Standard Model. Its best known quest, however, was the search for the famous Higgs boson particle predicted in the Standard Model (see page 90). Data collection began with the LHC's inaugural run in 2009, and each following run gradually narrowed down the energy range at which the particle might be found. On 4 July 2012, scientists announced that the Higgs boson had finally been identified in particle collisions with an energy of between 125 and 127 billion eV as theory predicted.

The LHC's other discoveries include several new particles and the creation of a new, superdense state of matter called a quark–gluon plasma, which exists at temperatures as high as 5.5 trillion °C (9.9 trillion °F). Upgrades to make the accelerator even more powerful will doubtless lead to further discoveries.

Main tunnel of the LHC

The Higgs boson

Imagine you've just been dropped into a big vat of treacle and are slowly sinking to the bottom. Then imagine that a ball-bearing – tiny, but with the same mass as yourself – is also dropped into the treacle. It sinks much faster, reaching the bottom before you, because it has a smaller cross-sectional area with which to create resistance in the treacle.

According to British physicist Peter Higgs, this is a good analogy for the process that gives particles their mass. They too have to wade through a medium of sorts, in this case an invisible one known as the Higgs field. The theory, developed by Higgs in 1964, predicts that just like the fundamental forces, the Higgs field is distributed across spacetime by a boson (see page 112).

The Higgs boson was finally discovered by the Large Hadron Collider in 2012, but scientists have yet to decipher many of its mysteries: in particular, the crucial question of why some particles are more massive than others.

Record of a collision giving
rise to a Higgs Boson particle
in the LHC's Compact Muon
Solenoid (CMS) experiment

Electromagnetic force

Alongside gravity, electromagnetism is the force that we experience most in our daily lives. Whether we're typing at a keyboard, turning the page in a book or opening a door, everything we do involves interactions between molecules and atoms that rely on the electromagnetic forces between charged nuclei and electrons (in particular, the outermost electrons in an atom, known as valence electrons, see page 60). Electromagnetic forces are responsible not only for the strong bonds between atoms in molecules and solid materials, but also for the weaker bonds that bind molecules more loosely in liquids and some gases. Our everyday lives are therefore built upon the foundations of quantum phenomena.

Like other fundamental forces, the electromagnetic force is carried by a gauge boson particle – in this case, photons of electromagnetic radiation. Because photons travel at the speed of light, the electromagnetic force can, in theory, have an almost infinite range.

Fine structures called lamellae on the feet of geckos allow them to use Van der Waals' forces (a weak electromagnetic attraction) to scale impossibly smooth surfaces.

Strong force

If the great range of the electromagnetic force makes it the long-distance runner of the Universe, the so-called 'strong force' is more like a 100-metre sprinter. It is incredibly powerful over the short scales of the atomic nucleus (distances of around one million billionths of a metre, or 10^{-15} metres, also known as a femtometre). In fact, at this scale, it's 137 times stronger than the electromagnetic force. This allows it to easily overcome the electromagnetic repulsion that is trying to drive positively charged protons apart within the nucleus.

On the scale of the nucleus, the strong force is carried between baryons through the exchange of messenger particles called pions (a type of meson, see page 94). The strong force is also what holds quarks together inside baryons and mesons. The force carriers exchanged between the quarks are gluons (of which there are eight types). The strength of the strong force is why we never see lone quarks outside the nucleus: they simply cannot escape its grip.

The strong force: binding the nucleus

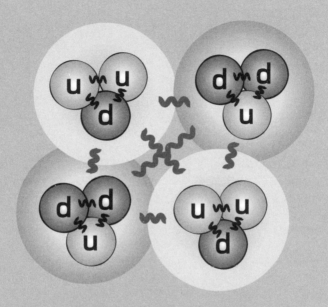

 Strong force carried between quarks by gluon particles

 Strong force transmitted between baryons by pion particles

Weak force

The weak force gets its name not because it is intrinsically weak, but because of its exceptionally short range. At scales of one billion billionths of a metre (10^{-18} metres), it is stronger than the electromagnetic force, but at just 10^{-17} metres, electromagnetism gets the upper hand.

The weak force plays the key role in beta decay, a type of radioactivity in which neutrons change spontaneously into protons and emit electrons. The reason for its short range is because neutrons in atomic nuclei must 'borrow' energy from the Universe in order to decay and emit one of the weak force's carriers (a charged W^+ or W^- boson, or a neutral Z boson). These bosons are far more massive than a neutron, and the cosmic debt collector doesn't loan this much energy out for long; hence, such bosons only last for one trillion trillionths of a second before being absorbed by another particle in order to repay the energy deficit. Beta decay is also a chiral interaction (see page 104): the preservation of 'handedness' helps determine which particular interactions can occur.

Common weak interactions

$$d \rightarrow u + W^-$$

Beta decay normally involves a down quark spontaneously changing to an up quark, transforming a neutron into a proton and releasing a W^- boson.

$$W^- \rightarrow e^- + \bar{\nu}_e$$

The W^- boson rapidly decays into an electron and an antineutrino.

$$c \rightarrow s + W^+$$

Inverse beta decay releases a W^+ boson, for instance when rare charm quarks spontaneously change into strange quarks.

$$W^+ \rightarrow e^+ + \nu_e$$

The W^+ boson decays into a positron (antielectron) and a neutrino.

$$e^- \rightarrow e^- + Z^0$$

Neutral-current interactions see electrons emit or absorb Z^0 bosons in high-energy environments such as particle accelerators.

$$Z^0 \rightarrow b + \bar{b}$$

The Z^0 rapidly decays into a fermion and its antiparticle, such as the bottom and anti-bottom quarks shown here.

Radioactivity

We're familiar with radioactive phenomena in modern life, ranging from natural radon gas to nuclear power, and from nuclear weapons to radiocarbon dating. In fact, radioactivity is a quantum physical process; no one can predict when an atom will undergo radioactive decay, because it is a random, spontaneous affair dictated by a probabilistic wave function.

Energy emitted by radioactive decay is released when a nucleus becomes unstable, usually because it has insufficient 'binding energy' to hold large numbers of protons and neutrons together against the electromagnetic repulsion of the like-charged protons. The atom's solution is to release some of these excess protons and neutrons through a process of alpha or beta decay, while also shedding excess, quantized 'gamma radiation'. The process transmutes the configuration of the nucleus, often into a different element. While it's impossible to predict the decay of an individual atom, when many radioactive atoms are present, it is possible to calculate the 'half life' – the time it takes for half the sample to decay.

The probabilistic nature of radioactive decay gives rise to a characteristic decay curve in which the quantity of radioactive parent atoms halves repeatedly over time.

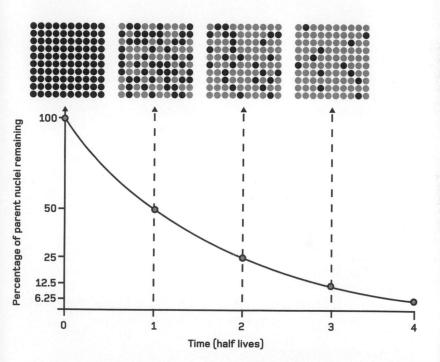

Alpha decay

Sometimes a particularly heavy element will release a cluster of two protons and two neutrons through radioactive decay. This particular combination of protons and neutrons is, in effect, a helium nucleus. In terms of radioactivity, however, it is known as an alpha particle. Removing these neutrons and protons transforms the parent atom into a different element. For example, the alpha decay of uranium-238 produces thorium-234.

Alpha decay is only possible thanks to one of quantum physics' more bizarre consequences. An alpha particle requires around 25 million eV to escape from the binding energy of a nucleus, but normally has between 4 and 9 million eV. In ordinary circumstances this should trap it within the nucleus, behind an energy wall known as a Coulomb barrier. However, all particles have an associated wave function and, if the alpha particle's wave function extends beyond the Coulomb barrier, there is a small probability of it appearing outside of the barrier. This phenomenon is known as quantum tunnelling (see page 170).

Alpha decay of Americium-241

The synthetic element Americium is a common component of smoke detectors. Alpha particles emitted during its decay ionize air and allow it to conduct electricity, completing a circuit. When smoke enters the detector, it blocks the passage of alpha particles and stops the current.

Helium nucleus
(alpha particle)
2 protons
2 neutrons

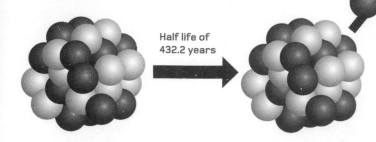

Half life of
432.2 years

Americium-241 nucleus
95 protons
146 neutrons

Neptunium-237 nucleus
93 protons
144 neutrons

Beta decay

As with alpha particles, the term beta particle is used to refer to a particle we've already met, but when it is emitted by radioactive decay. In this case, the particle is an electron (and sometimes the electron's antiparticle, a positron).

Beta decay involves the transformation of a neutron into a proton or, less commonly, a proton into a neutron. Each of these particles is made of three quarks: protons have two up and one down quark, while neutrons have one up and two down. Up and down quarks have slightly different energies and can morph into one another. In beta-minus decay, an up quark changes to a down quark, turning a proton into a neutron while releasing a negatively charged electron plus an antineutrino. The less common beta-plus decay turns a down quark into an up quark, changing a neutron into a proton and releasing a positively charged positron and a neutrino. Because the overall number of protons and neutrons remains the same, an atom undergoing beta decay does not change into a different element.

Two forms of beta decay

Beta-minus decay

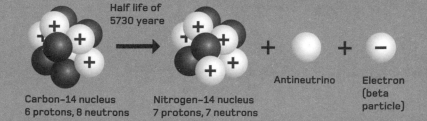

Half life of
5730 yeare

Carbon-14 nucleus
6 protons, 8 neutrons

Nitrogen-14 nucleus
7 protons, 7 neutrons

Antineutrino

Electron
(beta
particle)

Beta-plus decay

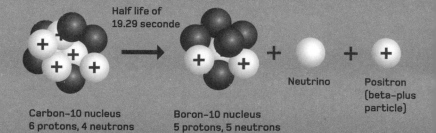

Half life of
19.29 seconde

Carbon-10 nucleus
6 protons, 4 neutrons

Boron-10 nucleus
5 protons, 5 neutrons

Neutrino

Positron
(beta-plus
particle)

Gamma decay

Following an atom's radioactive decay via either an alpha or beta particle, the parent nucleus is left in an excited state, with its component particles in a configuration that is less than optimal. Subsequently, the nucleus has to shed energy as these particles settle into their lowest-energy configuration (in a similar way to excited electrons, see page 68). Excited atomic nuclei shed this excess energy as photons of gamma radiation, the highest frequency of electromagnetic radiation. Gamma decay acts fast – usually within trillionths of a second after alpha or beta decay has taken place. For some nuclei, however, the process can take a little longer, perhaps one billionth of a second instead. Such comparatively long-lasting nuclei are termed 'metastable'.

The emitted gamma ray doesn't always escape into the world at large. Often it will collide with an orbiting electron in the atom's ground-state K-shell, giving that electron enough energy to escape from the atom entirely via the photoelectric effect.

Energy loss through gamma decay

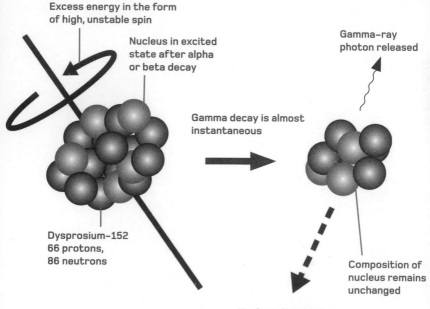

Excess energy in the form of high, unstable spin

Nucleus in excited state after alpha or beta decay

Gamma-ray photon released

Gamma decay is almost instantaneous

Dysprosium-152
66 protons,
86 neutrons

Composition of nucleus remains unchanged

Nucleus drops into a stable ground state, with lower spin

Virtual particles

Over previous pages we've mostly concerned ourselves with long-lasting, persistent particles. But quantum physics has another astonishing secret to reveal: much of nature is, in fact, held together by ghostly 'virtual particles' that only exist for the tiniest length of time, around 10^{-43} seconds. These particles can take almost any form, and fizz into existence from the vacuum energy of the Universe (see page 140). They defy the laws of conservation of mass and energy, but because they 'pay back' the energy they borrow in order to exist in such a short time, the Universe turns a metaphorical blind eye to them.

Many particle interactions depend on virtual particles: the photons that carry the electromagnetic force between electric charges are virtual, as are the gluons that hold quarks together, and the W and Z bosons that wield the weak force. The Van der Waals forces that bond molecules together, and even the electrostatic force attracting two metal plates in the Casimir effect (see opposite) also rely on virtual photons.

The Casimir effect is a weak force that arises between parallel metal plates separated by a very small distance in a vacuum. It arises because while virtual photons of many different wavelengths can pop in and out of existence in the space around the plates only short-wavelength ones can exist in the space between them.

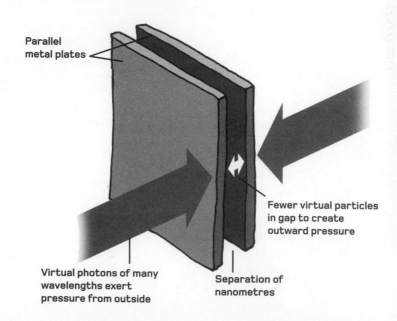

Parallel metal plates

Fewer virtual particles in gap to create outward pressure

Virtual photons of many wavelengths exert pressure from outside

Separation of nanometres

Lamb shift

During the 1930s, physicists realized that measurements of the binding energy of electrons in hydrogen atoms differed slightly from theoretical predictions. The problem was most noticeable in the excited P-shell (see page 60), where the theory was just 0.00003 per cent off, but even that was enough to cause consternation amongst atomic physicists.

It turned out that empty space itself was to blame. In quantum mechanics, space is never truly empty, but is filled with quantum fields that fizz with enough energy to produce countless virtual particles that appear and disappear within fractions of a second. Among these quantum fields is a superposition of random electromagnetic fields, each with its own associated virtual photon. These photons, popping in and out of existence, jostle the electron, pushing it around in random directions that create a shift in the electron's binding energy. The shift is named after Willis Lamb, whose experiments proved the theory, and ultimately led to the development of quantum electrodynamics.

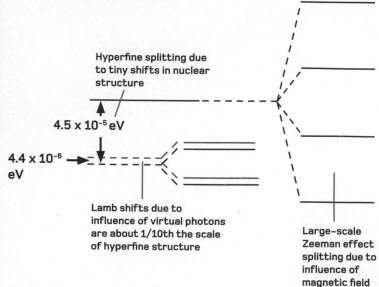

Willis Lamb's 1947 experiment measured three different types of spectral line splitting that arise when a beam of hydrogen atoms are exposed to microwave radiation in a magnetic field.

Hyperfine splitting due to tiny shifts in nuclear structure

4.5×10^{-5} eV

4.4×10^{-6} eV

Lamb shifts due to influence of virtual photons are about 1/10th the scale of hyperfine structure

Large-scale Zeeman effect splitting due to influence of magnetic field (see page 86)

Vacuum energy

Where does the energy to create short-lived virtual particles come from? The Universe is filled with a raw energy, known as vacuum energy, governed by wave functions that allow for the possibility that some of this energy will spontaneously turn itself into mass as virtual particle/antiparticle pairs (see page 136). This is possible only because of the inherent fuzziness of quantum physics, as described by the famous Heisenberg uncertainty principle (see page 172). In short, the principle makes it impossible to know the exact energy of the cosmos at any given time. If the Universe's accounting of its total energy is inherently askew, it's unlikely to miss those tiny quanta of energy borrowed by virtual particles for a short time.

Virtual particles are created everywhere all of the time. The Universe is constantly fizzing with their creation and destruction. Some scientists say this causes spacetime itself to fluctuate on the tiniest scales, forming a 'quantum foam'.

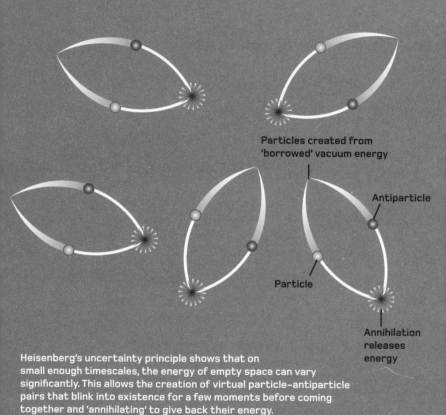

Particles created from 'borrowed' vacuum energy

Antiparticle

Particle

Annihilation releases energy

Heisenberg's uncertainty principle shows that on small enough timescales, the energy of empty space can vary significantly. This allows the creation of virtual particle–antiparticle pairs that blink into existence for a few moments before coming together and 'annihilating' to give back their energy.

The wave–particle world

Saying that particles also act as waves is one thing, but what does this mean in reality? The particle isn't a physical wave like a water wave on the ocean or a sound wave moving through the air. Instead, it's a description of the probability that a particle will have a given location or momentum when measured. This, in turn, produces many results that we interpret as particles acting like waves. For example, if you collect a large number of particles, such as a laser beam of photons or a beam of electrons fired from a hot electrode, then the spread of probabilities means that you can conduct experiments that show them acting en masse like waves, rather than discrete particles.

Don't worry if this is hard to picture – it is counterintuitive to how we see the everyday world. Yet the repercussions of 'wave–particle duality' are profound: concepts such as complementarity, uncertainty and decoherence (see pages 182, 172 and 176, respectively) completely change the way we think about reality.

Crystallography studies the small-scale structure of materials by analysing the way in which electrons are diffracted as they pass through gaps between atoms.

Probability wave function

Often represented by the Greek letter psi (ψ), a wave function offers a description of the different outcomes of a quantum system, and the probability that a particle will have a given solution. In practice, a particle's wavelike properties are never observed to be smeared out along a wave. Rather, peaks in the particle's wave function describe areas where it will is more likely to be appear: the taller the peak, the greater the likelihood. This implies that quantum physics is all about probabilities, which is why nothing is ever truly certain (at least not in the particle world). There's always a chance that a particle could be elsewhere, or have a different amount of momentum or energy.

How we interpret wave functions has far-reaching ramifications. The Copenhagen interpretation says that once a measurement is made, the wave function 'collapses' to a single solution. In contrast, the many-worlds theory (see page 286) predicts that every possible solution happens somewhere, each one in a parallel universe.

A simple probability wave function

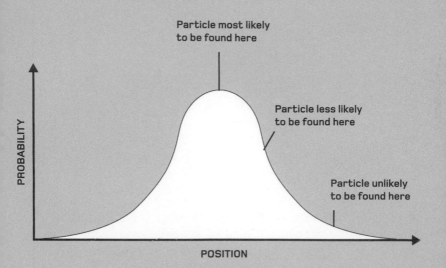

Applying Copenhagen

Although it's not the only way to interpret the meaning of quantum physics, the Copenhagen interpretation (see page 54) has proved the most popular over many decades. In some ways, it's also possibly the least imaginative interpretation, because it denies the idea of any deeper meaning to wave–particle duality: what we see is what we get.

The Copenhagen interpretation says that the wave function is the complete description of all the measurable properties of a particle and, conversely, that the properties of a particle are entirely based on the probabilities described by the wave function. This last point troubled Albert Einstein, prompting him famously to claim that 'God does not play dice'. The interpretation also states that a particle is not actually a wave and a particle at the same time, only that experiments designed to measure waves (such as Young's slits, see page 40) will see a wave, while experiments designed to measure particles will detect particles.

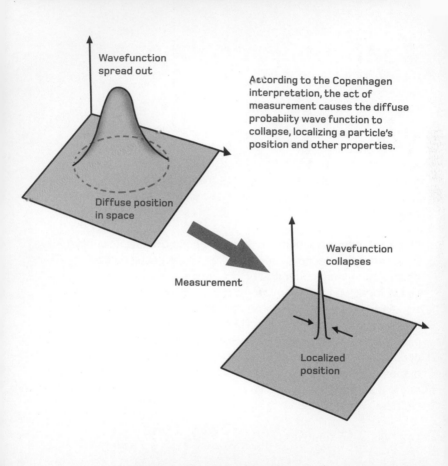

Wavefunction
spread out

According to the Copenhagen interpretation, the act of measurement causes the diffuse probabiity wave function to collapse, localizing a particle's position and other properties.

Diffuse position
in space

Measurement

Wavefunction
collapses

Localized
position

Quantum probabilities

What exactly are we referring to when we talk about 'probabilities' in a quantum context? Picture a wave function as a curve, or series of curves, on a graph like the one shown opposite. The x-axis is a measure of a particle's possible position, while the y-axis is a property known as amplitude, related to the probability that the particle will appear in a given position.

The most probable location is the highest peak on the wave function. As the amplitude decreases at greater distance, so too does the probability of the particle appearing at those locations. The sum of all these probabilities always equals 1, meaning that if the particle exists, it must be found somewhere along that wave function. So if you fire a beam of electrons through a narrow slit, most will follow the most probable path, but small percentages will be more broadly dispersed, making it seem as though the beam is a wave. This same mechanism explains how the quantum tunnelling vital to radioactive decay can take place (see page 170).

The wave function of a confined particle in a box can take on a number of different forms depending on the particle's energy, resulting in patterns of probability distribution similar to the harmonics of a vibrating violin string.

Low-energy particle

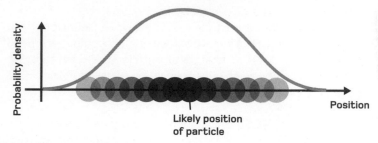

Likely position of particle

High-energy particle

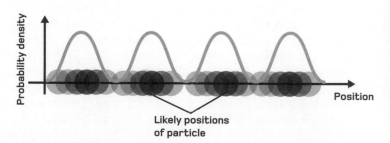

Likely positions of particle

The Born rule

The Born rule is a deceptively simple calculation of the probability that a particle exists at a given location along its wave function. In other words, if we make a measurement to see if a particle is at that location, then the Born rule (discovered by physicist Max Born in 1926) gives us the probability that we will indeed find it there.

The Born probability is simply the square of the amplitude of the wave function – that is, the height of the curve on a graphical representation of the wave function – at a particular location. The deceptive element of this simplicity arises from the fact that, if we probe a little deeper into the workings of quantum physics, we can't find any satisfactory reason *why* the square of the amplitude (multiplying the amplitude by itself) should give this probability. Given the Born rule's importance to quantum physics, this is a profound mystery, but the rule provides the link between theoretical predictions of quantum properties and our ability to experimentally measure them in the laboratory.

$$p(x,y,z) = | \psi(x,y,z, t_0) |^2$$

Probability density function   Wave function at time t_0

Quantum states

With many different quantum numbers in play on the subatomic scale, and wave functions to complicate things further, simply determining the properties of a quantum system could easily become problematic. Fortunately, however, a concept called the 'quantum state' offers a convenient way in which to package all the information about a quantum system, from its position and momentum to its quantum numbers.

However, because quantum physics is inherently probabilistic, the quantum state must be a distribution of *all* possible values for *all* of the aforementioned properties. Consider, for example, an alpha particle that is trapped in the so-called 'potential well' created by the Coulomb barrier around an atomic nucleus, with a wave function extending beyond the barrier (see page 168). The quantum state includes all outcomes of the particle's wave function, including both the possibility that it remains inside the potential well and that it tunnels out. This 'split personality' is at the heart of some of the strangest aspects of quantum physics.

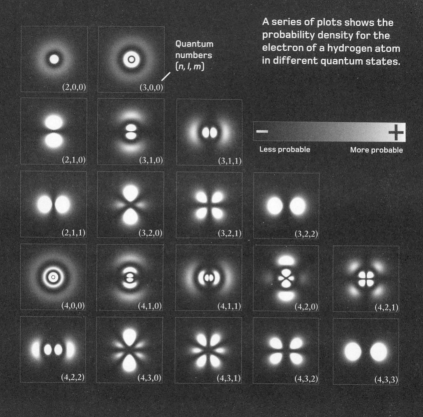

Quantum numbers (n, l, m)

A series of plots shows the probability density for the electron of a hydrogen atom in different quantum states.

Less probable   More probable

(2,0,0)   (3,0,0)

(2,1,0)   (3,1,0)   (3,1,1)

(2,1,1)   (3,2,0)   (3,2,1)   (3,2,2)

(4,0,0)   (4,1,0)   (4,1,1)   (4,2,0)   (4,2,1)

(4,2,2)   (4,3,0)   (4,3,1)   (4,3,2)   (4,3,3)

Quantum superpositions

The ability of a quantum state to have a distribution of possible outcomes is known as superposition. Think of each outcome having its own wave function, and all those wave functions overlapping or being superposed on top of each other. Alignments of the troughs and peaks cause those troughs and peaks to increase in amplitude, while misalignments cause the wave functions to cancel out. This pattern of constructive and destructive interference is exactly how everyday waves, such as sound waves, behave.

In quantum systems, however, the patterns of superposition are rather abstract. There is no actual physical wave to amplify or intensify, but instead superposed peaks increase the probability that, for example, a particle will be in a given position. Adding two or more wave functions together can therefore create a new quantum state. The double-slit experiment (see page 14) offers a good real-world example of quantum superposition at work.

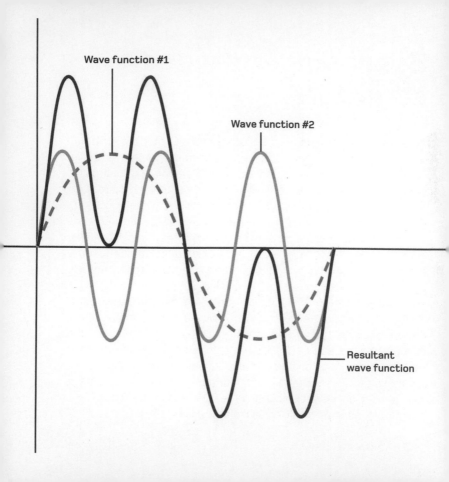

Schrödinger's wave equation

Many quantum mechanical concepts are so abstract that words and illustrations can never hope to describe them completely. Instead, they can only be fully explained using the language of mathematics. The most important mathematical description of all is offered by Schrödinger's wave equation. Developed by physicist Erwin Schrödinger in 1926, the equation was initially used to describe the quantum states of electrons in atoms. However, it can be adapted to describe quantum systems on any scale, up to the size of the Universe itself.

There are two versions of the equation; a time-independent form shown opposite (for non-moving particles), and a time-dependent one (used for a moving particle with a given location at a given time, see page 191). At the heart of the equation lies the concept of the wave function, also invented by Schrödinger. The equation, and, consequently, the Copenhagen interpretation, says that the wave function is the most complete description of a particle possible.

Time–independent Schrödinger equation in one dimension *x*

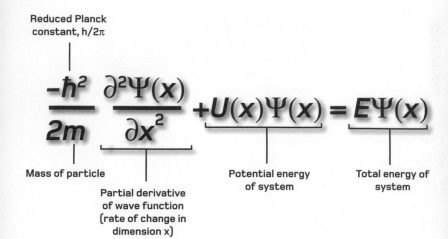

Reduced Planck constant, h/2π

$$\frac{-\hbar^2}{2m} \frac{\partial^2 \Psi(x)}{\partial x^2} + U(x)\Psi(x) = E\Psi(x)$$

Mass of particle

Partial derivative of wave function (rate of change in dimension x)

Potential energy of system

Total energy of system

Quantum harmonic oscillators

The classic example of a harmonic oscillator is a pendulum: assuming there's no frictional force such as air resistance to slow, or 'dampen', the pendulum's swing, it will oscillate with constant frequency and amplitude. Many physical phenomena involve vibrations, and physicists often use a mathematical model based on this kind of idealized motion to describe them.

The quantum harmonic oscillator is simply the quantum analogue to the classical models of the pendulum, and is useful for understanding a huge variety of quantum systems, such as a molecule with two atoms vibrating around fixed positions. In a quantum pendulum, the weight on the end of the wire could exist anywhere along the wave function, and since the energy of the ball is quantized (and the energy helps determine the frequency of the oscillation), this means quantum harmonic oscillators are also quantized. In other words, the wave function is only relevant at specific quantized energy levels.

Special relativity

Einstein's special theory of relativity describes the counterintuitive effects that happen when an object moves at 'relativistic' speeds, close to the speed of light. Einstein showed that the speed of light remains the same (299,800 kilometres or 186,000 miles per second) for all observers, no matter where they are or how fast they are travelling. In order to accommodate this remarkable fact, he found that a variety of other physical properties show strange behaviour.

For example, from the point of view of an outside observer, clocks run slower at relativistic speeds, while the dimensions of a moving object appear shorter. Because of $E = mc^2$ (see page 50), the mass of an object travelling ever closer to the speed of light also seems to increase, until the mass needed to reach light-speed itself becomes infinite. Relativity raises issues for quantum physics because special formulations of its equations are required to accurately describe what happens in the quantum world at close to the speed of light.

When astronauts spend time in space travelling at high speeds, time is actually running slightly more slowly for them as measured by an observer on Earth.

The Klein–Gordon equation

One of special relativity's golden rules is that nothing travels faster than the speed of light – neither physical objects nor information. Despite all the weirdness in quantum physics, this is one rule that cannot be broken, and it's one of the biggest tests that relativistic quantum theory has to face.

Schrödinger's wave equation is not relativistic – it deals with 'wave packets' (groups of superposed wave functions belonging to a particle) as standing waves with negligible velocities. Schrödinger himself developed a relativistic version of the equation initially, but found it inaccurate because it didn't incorporate spin. Subsequently, numerous physicists, including Swede Oskar Klein and Germany's Walter Gordon, revisited the relativistic version and showed how it could describe the relativistic motion of particles with zero spin, such as the Higgs boson. Fortunately, the Klein–Gordon equation, as it has since become known, shows that wave packets do not travel faster than light, preserving the cosmic speed limit.

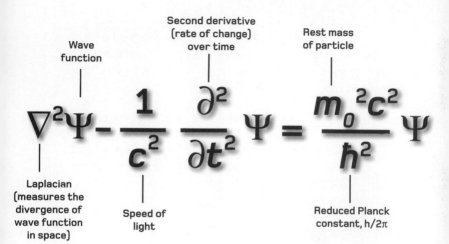

Wave function

Second derivative (rate of change) over time

Rest mass of particle

$$\nabla^2 \Psi - \frac{1}{c^2} \frac{\partial^2}{\partial t^2} \Psi = \frac{m_0^2 c^2}{\hbar^2} \Psi$$

Laplacian (measures the divergence of wave function in space)

Speed of light

Reduced Planck constant, h/2π

The Dirac equation

While the Klein–Gordon equation works for the very specific case of particles with zero spin, a much more adaptable equation is needed to deal with quantum effects across a wide range of particles at relativistic speeds.

It was shy British genius Paul Dirac who resolved this in 1928. His variation of the wave equation works for all particles with a spin of ½, meaning that, unlike the Klein–Gordon equation, it can be used to describe the energy levels of relativistic hydrogen atoms. However, as with the Klein–Gordon equation, some of the Dirac equation's solutions appeared to suggest particles with *negative* energy – a physical impossibility that doesn't make sense in the everyday world or even in our quantum picture of reality.

Fortunately, Dirac soon came up with a theory to explain this apparently nonsensical result. The solutions weren't literally particles with negative energy, he argued. Instead, they represented something else entirely: antimatter.

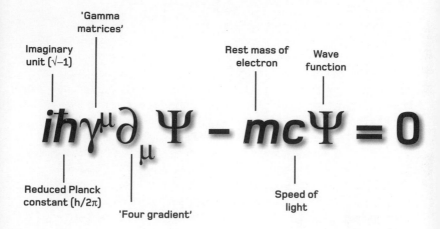

'Gamma matrices'

Imaginary unit ($\sqrt{-1}$)

Rest mass of electron

Wave function

$$i\hbar\gamma^{\mu}\partial_{\mu}\Psi - mc\Psi = 0$$

Reduced Planck constant ($h/2\pi$)

'Four gradient'

Speed of light

Antimatter

We've encountered antiparticles before (see, for example, page 127), but prior to the work of Paul Dirac in the late 1920s, the concept was completely unknown. The discovery of the first antimatter particle, the positron, came in 1932, and today we know that every particle in the Standard Model has a mirror-image antiparticle with the same mass but opposite charge (so if the charge on an electron is -1, then the charge on its antiparticle, a positron, is +1). Physicists have also proved that antiparticles can form their own 'antiatoms' and 'antimolecules'.

Put a particle and its antiparticle together and they annihilate each other in a flash of energy, producing a pair of high-energy photons in their stead. However, antimatter is extremely rare – at most the Large Hadron Collider could produce one-billionth of a gram of antimatter each year. Nobody knows why antimatter is so scarce, but it's a good job that it is – had there been equal amounts of matter and antimatter at the dawn of the Universe, their annihilation would have left a Universe filled only with photons.

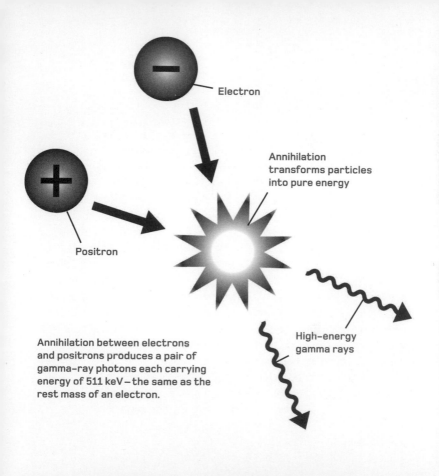

Electron

Annihilation transforms particles into pure energy

Positron

High-energy gamma rays

Annihilation between electrons and positrons produces a pair of gamma-ray photons each carrying energy of 511 keV — the same as the rest mass of an electron.

The Coulomb barrier

One intriguing consequence of the quantum nature of particles is their ability to be in one place at one moment, then reappear in a different place a moment later. This is what happens in alpha decay, when a helium nucleus 'tunnels' its way out of a larger nucleus. Before we explore quantum tunnelling in depth, however, we need a better understanding of the barrier through which a particle or nucleus has to tunnel.

In atoms, this barrier is called the Coulomb barrier. It's not a force field, but an electrostatic interaction between nuclei. At very close distances to the nucleus, it is attractive, but it becomes repulsive just a little further away. This makes it very good at keeping alpha particles in and very good at keeping other atomic nuclei out. The attractive Coulomb barrier that traps an alpha particle in the nucleus has an energy around 26 million eV, and should, according to classical physics, be insurmountable. Similarly, two nuclei trying to join together in a nuclear-fusion reaction require enough energy to overcome a repulsive Coulomb barrier.

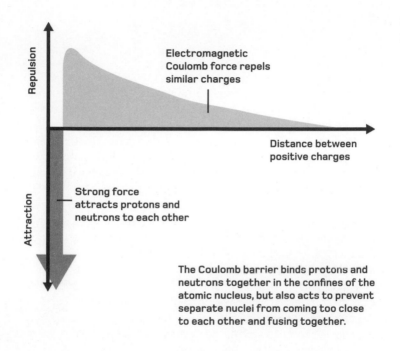

Repulsion

Attraction

Electromagnetic
Coulomb force repels
similar charges

Distance between
positive charges

Strong force
attracts protons and
neutrons to each other

The Coulomb barrier binds protons and
neutrons together in the confines of the
atomic nucleus, but also acts to prevent
separate nuclei from coming too close
to each other and fusing together.

Quantum tunnelling

There are two ways of overcoming the Coulomb barrier around an atomic nucleus. The first way is through sheer energy, but for a particle to attain a kinetic energy greater than the energy of the Coulomb barrier takes exceptionally high temperatures (tens of billions of degrees). Such high temperatures do not even exist in the centres of some low-mass stars, and yet hydrogen nuclei still merge to form helium nuclei and emit the energy that makes stars shine.

Instead, particles are more likely to resort to the second way of overcoming the barrier. The wave function of, say, an alpha particle or two atomic nuclei trying to fuse together can extend beyond the Coulomb barrier, meaning that there is a small chance that a particle can appear beyond (or penetrate through) the barrier. The size of the Coulomb barrier is different for each atomic nucleus so, depending on which nucleus is decaying, it may take anything from a few millionths of a second to many billions of years for an alpha particle to escape.

Classical model of alpha decay

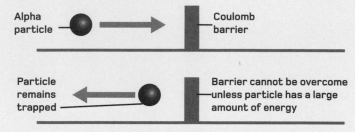

Alpha particle → **Coulomb barrier**

Particle remains trapped ← **Barrier cannot be overcome unless particle has a large amount of energy**

Quantum model

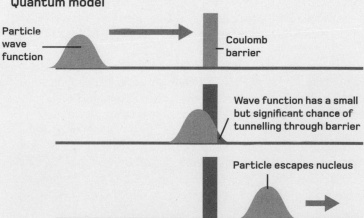

Particle wave function → **Coulomb barrier**

Wave function has a small but significant chance of tunnelling through barrier

Particle escapes nucleus →

Heisenberg's uncertainty principle

The fuzziness of quantum physics isn't only found in the wave function, but is also apparent in the properties we can measure in a particle. Werner Heisenberg stumbled upon this problem, realizing that it is impossible to be certain of both the momentum and the position of a particle at any one instant: the more accurately we know the value of one of these properties, the less accurately we know the other. Heisenberg called this his 'uncertainty principle'. It has nothing to do with errors in experimental apparatus or method but is, instead, a result arising from the behaviour of a particle's wave function. The more you can pin down a particle's location, the more tightly bunched the wave function (which determines the particle's position) becomes. However, this has the consequence of making the wave function's wavelength (which determines a particle's momentum) less precise. Conversely, the more accurately the wavelength is known, the more widely distributed the wave function is, giving the particle a greater probability of existing in many different locations.

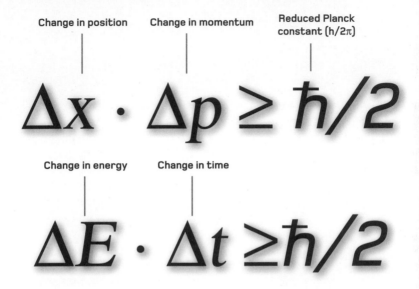

Change in position   Change in momentum   Reduced Planck
                                            constant (h/2π)

$$\Delta x \cdot \Delta p \geq \hbar/2$$

Change in energy   Change in time

$$\Delta E \cdot \Delta t \geq \hbar/2$$

The uncertainty principle states that certain complimentary
pairs of quantum properties (position and momentum, or
energy and time) cannot both be determined with perfect
accuracy in the same measurement.

Uncertainty in action

Heisenberg's uncertainty principle has some fascinating real-life uses. Magnetic resonance imaging (MRI), for example, relies on the relationship between the frequency of a radio-wavelength photon and its momentum. The uncertainty principle says we can't know a photon's frequency and its position at the same time, so in a burst of radio photons whose positions are known, frequencies will be uncertain and spread across a wide range — perfect for probing different parts of the human body.

The principle also has some other fundamental consequences. It dictates the size of electron orbitals, and hence the size of atoms themselves. It also explains why the oppositely charged electrons and protons in atoms are not attracted to each other: moving closer to the nucleus would pin down an electron's position so tightly that the uncertainty in its momentum would be huge. Quantum tunnelling is another manifestation of the uncertainty principle, while virtual particles (see page 136) can only exist because of uncertainty in the exact energy of space.

Uncertainty and the wave function

The more accurately an object's de Broglie wavelength (related to its momentum) is determined, the less accurately its location can be pinned down.

Constraining the object's position leads to greater uncertainty in its wavelength and momentum – a probem that gets worse the more tightly the position is defined.

Quantum decoherence

If we want to measure a single property of a quantum system – say the energy of an electron or the position of a proton – we can do so with great accuracy. When we make these measurements, the wave function is said to collapse. This is a misnomer, however: a particle doesn't actually lose its wavelike properties when a measurement is made. A better way of describing what happens is to say that quantum information leaks out of the system. This effect, called decoherence, was outlined in 1970 by Heinz-Dieter Zeh.

Zeh suggested that when the wave function of a measuring device comes into contact with that of a particle, it creates interference that causes the particle wave function to decay, allowing precise measurement. In some ways, this is the opposite of superposition: waves are scrambled by destructive interference rather than strengthened by constructive overlapping. Because larger objects are in constant contact with the wave functions of their environment, they decay much faster, explaining why quantum 'fuzziness' isn't apparent at everyday scales.

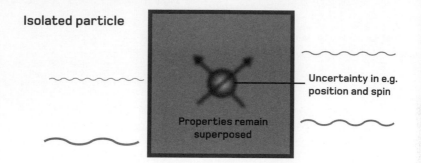

Isolated particle

Uncertainty in e.g. position and spin

Properties remain superposed

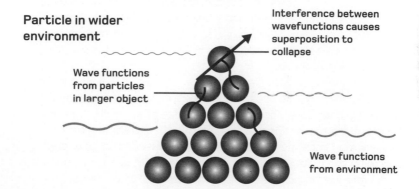

Particle in wider environment

Interference between wavefunctions causes superposition to collapse

Wave functions from particles in larger object

Wave functions from environment

Schrödinger's cat

Nothing lays bare the weirdness of quantum physics so well as Schrödinger's cat. Erwin Schrödinger created his famous feline thought experiment in 1935, to demonstrate what he saw as the absurdity of the Copenhagen interpretation (see page 54). Illustrated opposite, the experiment imagines a way of putting a macrocopic system (a cat that may be alive or dead) into a state of quantum superposition. To a person outside the box, there is no way of knowing whether and when a radioactive substance has decayed: therefore, Schrödinger said, Copenhagen implies the cat is both alive *and* dead at the same time until the box is opened.

Schrödinger argued that such a situation was intuitively absurd, but he really only addressed an extreme reading of Copenhagen in which a *conscious* observer is the cause of the wave function's collapse. Niels Bohr, for one, argued for a broader definition of observation in terms of interaction with macroscopic systems including the Geiger counter and the cat itself, which would therefore trigger collapse long before the box was opened.

Imagine locking a cat inside a box with no way to see in. A small quantity of radioactive material, placed in the box along with the cat, has a known probability of decaying in a given time frame, governed by quantum rules. If the decay event happens, a Geiger counter detects the radiation, and triggers the release of a poison that kills the cat.

Testing Schrödinger's cat

No cats were harmed in the making of this book, and indeed a real-world enaction of Schrödinger's thought experiment would serve no practical purpose since the entire point is that events in the sealed box remain undetermined until an outcome is observed. But scientists have tested the principle behind it in other ways and shown that, despite Schrödinger's scepticism, the Copenhagen interpretation does indeed reflect what happens.

The trick is to place a quantum system in a state of superposition of wave functions. In the cat experiment, these belong to the different outcomes of the cat being alive or dead. No one has succeeded in superposing such large quantum systems in reality, but in experiments photons, beryllium atoms and even a vibrating tuning fork composed of ten trillion atoms have all been superposed in oscillating states that cause them to act as though they are in two places at once. The experiments prove what we struggle to embrace: that on the smallest scales the Universe really is probabilistic.

Theory and observation both support the idea that wave functions really do remain superposed on quantum scales until they are observed.

The challenge is therefore to explain why the wave function then resolves itself, and why uncertainty doesn't really apply to cats and other large-scale objects.

Complementarity

To Niels Bohr, architect of the Copenhagen interpretation, the wave and particle aspects of nature were two sides of the same coin that complement one another. If a measurement observes an electron acting like a wave, that's because the experiment is set up to detect a wave (and likewise if it detects a particle). The act of measurement causes the electron's wave function to 'collapse' into the wave solution. At the same time, we cannot know the full properties of a particle without considering both its wave and particle natures, but the Heisenberg uncertainty principle puts this complete knowledge out of reach. The complementary nature of particles is both a blessing and a curse: it allows them to do extraordinary things and for the building blocks of the Universe around us to form and function. At the same time, however, it creates a sense of frustration that on a particle-by-particle basis, the quantum world is entirely random and not deterministic. For early 20th-century physicists who preferred the Universe to run like clockwork, this was a hard pill to swallow.

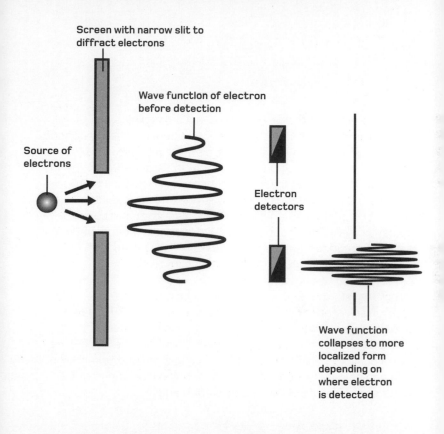

Screen with narrow slit to diffract electrons

Wave function of electron before detection

Source of electrons

Electron detectors

Wave function collapses to more localized form depending on where electron is detected

Quantum mathematics

Like all physics, the study of the quantum realm is an inherently mathematical subject. A scientific theory cannot rely on making statements of fact alone. It must also have a mathematical foundation that both explains observations and makes predictions. As we have seen, however, quantum physics is radically different to classical physics, and so it requires its own brand of mathematics and its own unique equations. Concepts that seem counterintuitive from a conceptual standpoint suddenly make more sense when viewed with a little more understanding of the maths involved.

Unfortunately, the mathematics of quantum mechanics is hard – it took some of the best scientific minds of the 1920s and 1930s to invent the systems required to make sense of it all. This book cannot hope to explain everything but, nevertheless, even a cursory look at some of the techniques and terminology used can help us to quantify quantum behaviour and provide deeper insights into what's really happening.

'Shut up and calculate!'

N. David Mermin

What is a matrix?

A common mathematical tool used in a wide variety of applications, including quantum physics, a matrix is simply a table of numbers arranged in rows and columns. The examples shown opposite are square, but equal numbers of rows and columns are not compulsory. Each number within the matrix is called an 'element'.

The advantage of a matrix is that it allows you to perform mathematical functions on each element sequentially. For example, matrices of equal dimensions can be added to, or subtracted from, one another by adding or subtracting the corresponding properties in each matrix (see opposite, above).

Multiplying is a little different, and requires the number of columns in the first matrix to be the same as the number of rows in the second. Each row in the first matrix then multiplies by each column in the second matrix and the products of those multiplications are added together (see opposite, below).

Matrix addition:

$$\begin{bmatrix} 1 & 0 & 2 \\ 2 & 1 & 3 \\ 3 & 2 & 1 \end{bmatrix} + \begin{bmatrix} 2 & 1 & 3 \\ 2 & 3 & 1 \\ 1 & 3 & 2 \end{bmatrix} = \begin{bmatrix} 3 & 1 & 5 \\ 4 & 4 & 4 \\ 4 & 5 & 3 \end{bmatrix}$$

Matrix multiplication:

$$\begin{bmatrix} 1 & 2 \\ 3 & 4 \end{bmatrix} \times \begin{bmatrix} 5 & 6 \\ 7 & 8 \end{bmatrix} = \begin{bmatrix} (1\times5)+(2\times7) & (1\times6)+(2\times8) \\ (3\times5)+(4\times7) & (3\times6)+(4\times8) \end{bmatrix}$$

$$= \begin{bmatrix} 19 & 22 \\ 43 & 50 \end{bmatrix}$$

Matrix mechanics

In the 1920s, quantum physicists struggled to conjure up a mathematical description of the strange wave-particle duality they were observing. A solution came from Max Born, who developed an idea from Werner Heisenberg that electron orbits were best described by harmonic waves. Heisenberg calculated the quantum jumps of the electrons through cumbersome equations that contained a lot of multiplication. Born realized that these sequences of multiplications could be much better described in matrices, where multiplying each matrix element helped calculate the electrons' spectral lines given their energies.

Yet, at the time, Born's 'matrix mechanics' approach proved quite unpopular. Matrices were considered an oddity of pure maths by most physicists in the 1920s, and so this seemed a strangely abstract way of depicting electron orbits. Consequently, Schrödinger's wave equation (see page 156) remained the more popular means of describing the quantum behaviour of particles.

Friedrich Hund, Werner Heisenberg and Max Born, the three founders of the matrix mechanics approach to quantum physics, reunited at a 1966

Wave mechanics

Schrödinger's wave equation takes a different tack to matrix mechanics, by turning the wavelike properties of a particle into an equation that describes their distribution. In essence, the equation describes how the quantum state of a system, defined by its wave function, changes over time. The Schrödinger equation is described as wave mechanics – the motion of wavelike matter. In that sense, it offers a quantum analogue to Newton's classical second law of motion: force equals mass times acceleration ($f = ma$).

There are several versions of the Schrödinger wave equation. The most common is the time-dependent equation, shown opposite. Here, i is the square root of -1 (an 'imaginary' number that does not exist as a real number, but is vital to the solution of some equations), $\hbar$ is the Planck constant divided by 2π, Ψ is the wave function (see page 144) and $\hat{H}$ is the Hamiltonian operator, which helps to describe the total energy of a quantum system (see page 198).

The time–dependent Schrödinger equation
for position vector *r* at time *t*

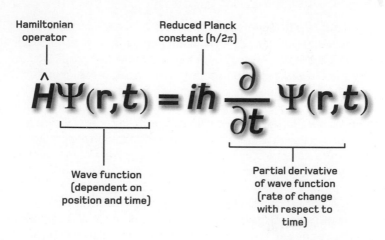

Hamiltonian
operator

Reduced Planck
constant (h/2π)

$$\hat{H}\Psi(r,t) = i\hbar \frac{\partial}{\partial t} \Psi(r,t)$$

Wave function
(dependent on
position and time)

Partial derivative
of wave function
(rate of change
with respect to
time)

Hilbert space

A vector is a mathematical quantity, such as velocity or acceleration, that has both a magnitude or strength, and a direction. We can draw it on a two-dimensional graph with x and y axes in what mathematicians call two-dimensional flat or Euclidean space. But suppose we want to measure vectors in an infinite (or arbitrary) number of dimensions, with an infinite (or arbitrary) number of coordinates, not just x and y? Such a structure is called Hilbert space, named after physicist David Hilbert. You can measure distances and angles within it just as you can in Euclidean flatland.

Hilbert space is important to quantum mechanics, where the vectors of classical mechanics are replaced by particles whose wave functions can have an infinite number of solutions. Both wave functions and solutions to the wave equation can be visualized in Hilbert space. What's more, string theory, a potential 'theory of everything' (see page 252), also appears to describe the Universe as a Hilbert space with up to a dozen dimensions.

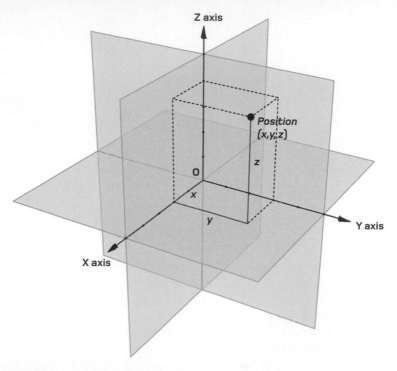

Three-dimensional space, in which positions can be defined in terms of coordinates along three perpendicular axes, is a familiar example of a Hilbert space.

Transformation theory

Although they arrive at the same physical consequences, matrix mechanics and the Schrödinger wave equation (see pages 188 and 156) represent very different versions of quantum mechanics. So when Paul Dirac (opposite) developed his transformation theory, not only showing how matrix mechanics and the Schrödinger equation are equivalent, but unifying them under a single mathematical umbrella, it was a major achievement.

Both matrix mechanics and the Schrödinger equation are descriptions of how quantum states evolve. Dirac unified the two models by depicting them as 'transformations' in Hilbert space. When a quantum state changes over time, the vector that describes its wave function is effectively moving between different locations in Hilbert space – a movement that Dirac described as the transformation. Matrix mechanics and the Schrödinger equation are just two different ways of describing such a transformation.

Quantum operators

In classical Newtonian physics, properties such as energy, velocity, momentum and position are given by real, defined numbers: a car travelling at a velocity of 40 kilometres per hour (25 mph) with a kinetic energy of 35,000 joules can be located at specific coordinates along every step of its journey.

But as we have seen through the wavelike nature of particles and Heisenberg's uncertainty principle, it is impossible to give exact values to such properties on the scale of atoms and particles. Instead, in quantum mechanics these properties are defined as 'operators' that form the basis of the mathematical language of quantum mechanics. Operators describe functions that transform one set of quantum states into another set. For example, the Hamiltonian operator (see page 198) is the operator that describes all possible outcomes when the total energy of a quantum system is measured. Similarly, the position operator describes the sum of possible outcomes when we attempt to measure the position of a particle.

Observable phenomenon		Operator Symbol
Name	Symbol	
Position	r	$\hat{r}$
Momentum	p	$\hat{p}$
Kinetic energy	T	$\hat{T}$
Potential energy	$V(r)$	$\hat{V}(r)$
Total energy	E	$\hat{H}$
Angular momentum (in x, y, z directions)	$l_{x,y,z}$	$\hat{l}_{x,y,z}$

Hamiltonian operator

One of the most important operators in quantum mechanics is the Hamiltonian. It describes the set of all possible outcomes when measuring the total energy of a quantum system. For a single particle, this is essentially the sum of the operators describing the particle's kinetic energy (derived from motion and mass), and potential energy (derived from its position in a force field). However the Hamiltonian can also be used when describing energy levels embedded within a system, such as those of electrons orbiting around an atomic nucleus.

Named after 19th-century Irish physicist William Hamilton, the operator can vary depending on the number of particles in the system being measured. It plays a major role in the time-dependent Schrödinger wave equation (see page 190), where it instigates evolution of the wave function over time. The eigenvalues arising from possible solutions to the equation (see page 204) correspond to the energy levels associated with those solutions.

Path integral formulation

As we've seen, the secret behind electron diffraction in the double-slit experiment (see page 40) is a particle's ability to display wavelike behaviour: a wave can pass through both slits, whereas in classical physics a particle's trajectory can pass through only one. The particle's wave function is therefore a probability distribution, describing all the different possible trajectories that the particle could take.

However, given the uncertainty in knowing which path a wavelike particle takes to get from A to B, how can we determine the most likely trajectory? The possible routes are virtually infinite, including those that go across the Universe and back. It was Richard Feynman, one of the greatest minds of the 20th century, who came up with a technique to calculate the most likely path of a particle, based on ideas initially developed by Paul Dirac. Feynman's 'path integral formulation' takes the sum of probabilities from the wave function for *every* possible path and mathematically combines or integrates them to find the most likely path.

A particle can take many different paths in moving from point A to point B in a given time. The path integral approach involves combining them all to find the most likely route.

Feynman diagrams

Why bother with complicated equations when you can just draw a picture? That sounds flippant, but it's essentially the diagrammatic approach pioneered by Richard Feynman when representing quantum interactions. A Feynman diagram can show particles represented as straight lines converging on a point called a vertex, where they interact. The interaction involves the exchange of a gauge boson – a photon, gluon, W^+ or W^- boson depending on which quantum field dominates the interaction. The gauge bosons are depicted by wiggly lines, and on the other side of the interaction the particles resulting from the interaction continue on their way (opposite, above).

Sometimes, a Feynman diagram can show the interaction between a photon and a single particle, such as an electron. In this case, the straight line of the electron and the wiggly line of the photon meet at the vertex. The particle absorbs the photon and is momentarily excited, represented as a single horizontal line, before re-emitting the photon (opposite, below).

Simple interaction between fermions

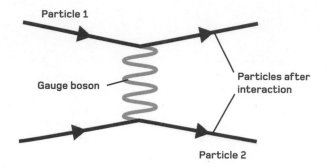

Electron excitation

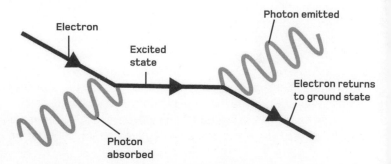

Eigenfunctions

Usually, when an operator acts on a wave function, it transforms the shape of the wave function and therefore the probability of various outcomes. But this isn't always the case; sometimes the operator creates multiple versions of the wave function instead. In this case, each of the resulting wave functions is called an 'eigenfunction' (a hybrid German-English word essentially meaning 'its own function').

A common example is the Hamiltonian operator describing the total energy of a quantum system. When applied to a wave function, it typically produces meaningful solutions of the Schrödinger equation only for certain discrete values of energy. Each of these possible energy states is known as an eigenvalue, and the wave function associated with it is its eigenfunction. In the broader case of any operator Q that produces quantized, discrete outcomes, the multiple outcomes created by an operator become eigenvalues. Each can be associated with an index i – a new quantum number that helps define the system.

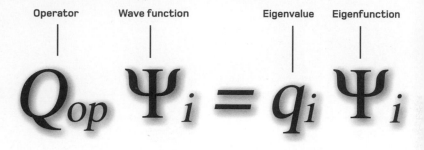

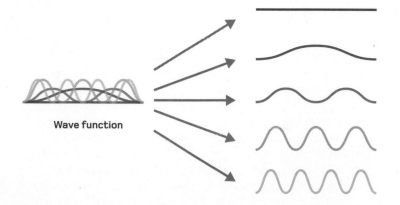

The correspondence principle

Quantum mechanics deals with the physics of the very small and, as we've seen, quantum behaviour is very different to the macroscopic behaviour in our everyday world dominated by classical physics. But there must come a point in scale at which quantum and classical behaviour overlap. At this point, calculations conducted using quantum mathematics must correspond to the results of classical mathematics.

Hence the 'correspondence principle', developed by Niels Bohr, which states that when quantum systems (or rather their quantum numbers) become sufficiently large, they must approximate to classical mathematics. This principle is very useful in determining which quantum mechanical models have any basis in reality. In fact, the concept serves well throughout science, dictating that any new theory must also be able to explain the results of any old theory that it replaces. For example, in order to succeed, Einstein's general relativity had to be able to explain Newton's laws of gravity in new terms, while matching their accurate predictions.

A quantum harmonic oscillator (see page 158) differs starkly from classical models in its ground state, but starts to resemble them much more closely in its higher states of oscillation.

Ground state wave function

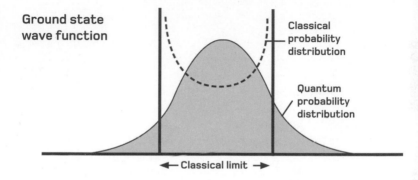

Classical probability distribution

Quantum probability distribution

← Classical limit →

Higher state wave function

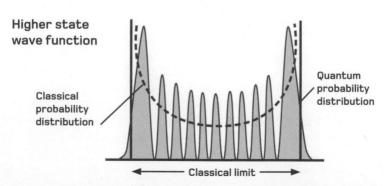

Classical probability distribution

Quantum probability distribution

← Classical limit →

Limits of the quantum realm

It's often said that quantum mechanics is the physics of very small things, and we know we don't see quantum mechanical behaviour in the large-scale everyday world; the position and momentum of people and cars and buildings all seem very precise to us. So at what point does the correspondence principle take hold? What is the largest piece of matter that can exhibit quantum mechanical properties?

One of the most popular tests for quantum properties is the double-slit experiment, which tests wave–particle duality. In theory, we could simply fire bigger and bigger chunks of matter at the slits and see when the interference fringes disappear. The largest molecules so far shown to produce wavelike behaviour are carbon 'buckyballs' about a nanometre (a billionth of a metre) wide. How much larger we can go before quantum behaviour breaks down is still unknown, but tests with viruses, which can be several hundred nanometres long, have been proposed. If they could show wave–particle duality, it would be truly remarkable.

Football-shaped carbon molecules known as buckyballs have many remarkable properties, not least of which is that they are the largest objects so far shown to display wave–particle duality.

Perturbation theory

While physicists can calculate solutions to Hamiltonian operators corresponding to, say, a quantum harmonic oscillator or the energy levels of a hydrogen atom with great precision, these are pretty idealized scenarios. Hamiltonian operators become rapidly more troublesome to solve exactly with the more complex Schrödinger equation.

Perturbation theory is a trick that physicists use to get around this problem. The idea is to start with a simple quantum system, such as a hydrogen atom, and then add a Hamiltonian operator that 'perturbs' it, tweaking it incrementally. The result is a system for which we already know the solution, with additional small corrections or modifications to quantum properties, such as energy levels. This is the method used to calculate variations in energy levels caused by phenomena such as the Stark effect (see opposite). Simple systems act as touchstones, allowing physicists to take known solutions and use them to explore the solutions to more complex ones.

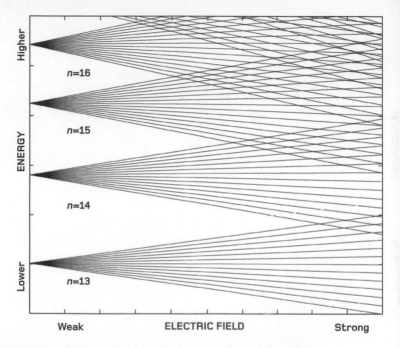

Application of an electric field splits the spectral lines of hydrogen (indicated by principal quantum number *n*) into numerous, sometimes overlapping, sublevels. The phenomenon, known as the Stark effect, is best modelled with perturbation theory.

The Universe

The greatest mystery in all of science is the origin of the Universe itself. How did it begin? How has it evolved since the dawn of time? Was there anything before it and will anything come after it? Are there other universes, and what does this all mean for life on Earth?

Throughout the ages, myth, superstition and religion have attempted to provide answers to these questions, but it is only relatively recently that science has been able to make a meaningful contribution to the discussion. Our telescopes probe ever deeper into the Universe, while our microscopes peer closer into the subatomic world in which quantum physics reigns. In doing so, they have started to provide answers, often involving quantum-level effects, that nevertheless wield considerable influence on the largest scales of the Universe. Indeed, it may be that the whimsical nature of wavefronts and quantum fluctuations decided the shape our cosmos would take before the first fraction of a second had even passed.

The Big Bang

The orgins of the Big Bang theory lie in the fact that space itself is expanding (see page 228). If the Universe is getting bigger now, then it must have been smaller in the past. Extrapolating backwards, scientists believe that everything in today's Universe originated at a single point in space around 13.81 billion years ago.

Contrary to popular belief, however, the Big Bang does not explain what *caused* the birth of the Universe, only what happened immediately afterwards. The Big Bang itself is shrouded in mystery: for the first trillion trillion trillionth (10^{-36}) of a second conditions were so intense (all the energy in the Universe packed into a volume the size of an apple with a temperature in excess of 10^{32}°C) that our understanding of physics utterly falls apart. This earliest fraction of a second was dominated by the unknown rules of quantum gravity that marry quantum physics with Einstein's relativity (see page 48). Even the Big Bang itself may have been the result of a random quantum fluctuation that allowed the mass-energy of the Universe to pop into existence in a similar way to today's virtual particles (see page 136).

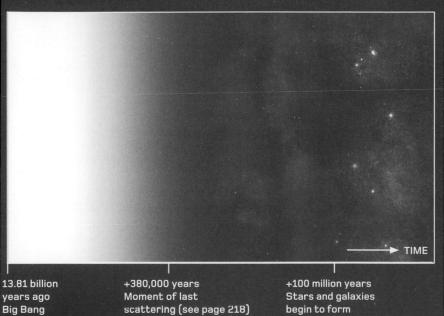

→ TIME

13.81 billion
years ago
Big Bang

+380,000 years
Moment of last
scattering (see page 218)

+100 million years
Stars and galaxies
begin to form

The Big Bang theory explains how the Universe began as a ball
of matter in a hot, dense state where the laws of quantum
physics dominated. As a result, quantum theory plays a vital
role in explaining many aspects of today's cosmos.

Quantum fluctuations

In the first tiny fractions of a second of the Big Bang, there were no atoms, no protons, no quarks, no particles of matter at all, just pure energy. Ruled by quantum gravity, this energy was subject to myriad quantum fluctuations. Virtual particles popped in and out of existence everywhere, resulting in stark variations in energy density across the tiny fledgling Universe. These fluctuations had such a lasting influence that they dictated the large-scale distribution of matter, in the form of galaxy clusters and superclusters, that we see in the Universe today.

Einstein's famous equation $E = mc^2$ shows that mass and energy are two sides of the same coin and, as the Universe expanded and cooled, much of the original raw energy transformed into matter. Regions in which the energy density was greatest naturally had a greater density of matter. Without those early quantum fluctuations, energy and matter would be spread much more evenly across space, with particles scattered so thinly that no stars, planets or galaxies would be able to form.

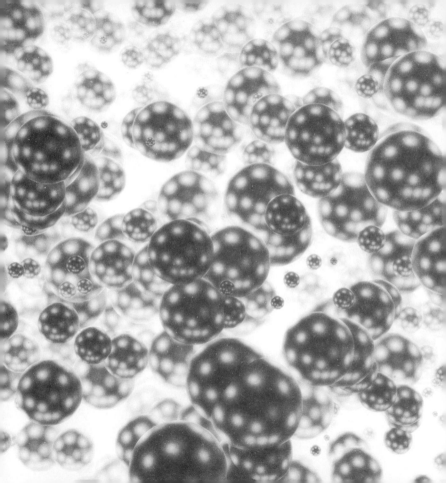

Cosmic microwave background radiation

For 380,000 years after the Big Bang, the Universe was a sea of plasma, a state of matter within which atomic nuclei and free-floating electrons formed a kind of electrically charged 'soup'. Photons of light attempting to travel through the Universe would continually scatter off the electrons, ricocheting like light trapped in fog. However, as the Universe expanded, it cooled, and the temperature dropped sufficiently for electrons to be captured in orbit around atomic nuclei. As they were absorbed, complete atoms formed for the first time (mostly hydrogen, with small proportions of helium and lithium), and photons were at last able to travel unhindered through space.

Today, we detect photons from this event, called the moment of last scattering, in the cosmic microwave background radiation (CMBR) – short-wavelength radio waves that can be detected from all parts of the sky. Tiny temperature variations within it are a record of the quantum fluctuations that the Universe experienced in the very earliest moments of time.

This detailed map of the CMBR from NASA's WMAP satellite shows variations in temperature and density that must have been seeded in the moments after the Big Bang by quantum-level variations in the infant Universe.

The origin of galaxies

The temperature variations seen in the cosmic microwave background, resulting from quantum fluctuations just after the Big Bang, went on to form the seeds of galaxies. The fluctuations created differences in the energy density of parts of the infant Universe and, as the Universe cooled, the raw energy condensed into matter distributed unevenly in space.

Areas that were denser would have had stronger gravity. Over time these began to pull other matter towards them, generating a sort of 'cosmic web', with long filaments of matter stretching for millions of light years around much larger voids. The filaments of the cosmic web mirror the pattern that we see today in the distribution of galaxy clusters, each containing hundreds to thousands of galaxies. Arrangements of many clusters form chains or walls of galaxies that are the biggest structures in the known Universe. Everything that we see around us in the Universe, including these, is a result of initial quantum fluctuations that were frozen in place as the Universe expanded.

A computer simulation shows the formation of galaxies and galaxy clusters around knots of dark matter in the early Universe. The weblike distribution of dark matter is directly attributed to quantum fluctuations in the moments after the Big Bang.

The horizon problem

The furthest distance into the Universe that our telescopes can see is the 'cosmic horizon'. Light from anything beyond a certain distance simply hasn't had enough time to reach us since the Big Bang. Astronomers calculate the distance to the cosmic horizon at 46.5 billion light years, meaning that Earth lies in the centre of a spherical volume 93 billion light years across, the limit of our observable Universe. The reason that we can see so far, despite the Universe being 13.81 billion years old, is because of cosmic expansion: light left the most distant visible galaxies 13 billion years ago, but they have since been carried even further from us.

This raises an important question: light might have had time to reach us, but neither it nor any other kind of information can have reached the opposite horizon. Yet the Universe looks remarkably similar in all directions, more than can be explained if they have not had contact since the instant of the Big Bang itself. Solving this 'horizon problem' required a new theory that would transform our understanding of the first second of cosmic history.

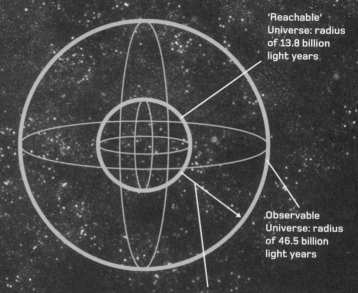

'Reachable' Universe: radius of 13.8 billion light years

Observable Universe: radius of 46.5 billion light years

Spacetime similar to our own stretching far beyond the limits of our observable Universe

Cosmic expansion has carried the most distant galaxies much further away while their light has been travelling to reach us.

Inflation

In 1980, cosmologist Alan Guth came up with an ingenious solution to the horizon problem. He decided that if opposite sides of the visible Universe looked like they had once been in so-called 'causal contact' (that is, close enough for events in one part of space to affect the other), then maybe that was because they had. Perhaps somehow the seeds of our observable Universe had indeed stayed together for a little longer than had previously been suspected?

Guth's theory, called inflation, proposed that in the very first fractions of a second of creation, 'our' Universe remained small enough for causal contact. But then, just 10^{-33} seconds after the Big Bang, a sudden burst of accelerating energy called inflation caused the Universe to balloon dramatically in size, pushing areas that were once in contact so far apart that light could never again reach from one side to the other. Inflation lasted for an instant, but it played a key role in magnifying the quantum fluctuations of the infant Universe (see page 216).

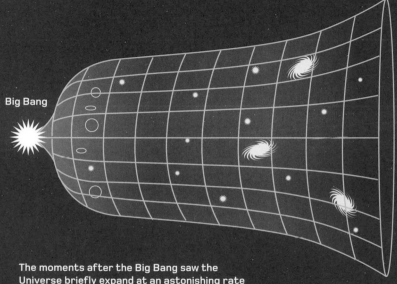

TIME

Big Bang

The moments after the Big Bang saw the
Universe briefly expand at an astonishing rate
before slowing down to its present, more sedate
pace. Inflation is thought to have blown up the
early Universe by some 26 orders of magnitude.

Eternal inflation

The theory of inflation still raises unanswered questions. Nobody truly understands why it began or, indeed, what caused it to stop. Indeed, several cosmologists have proposed that, in some parts of the Universe, inflation never stopped.

One idea is that during inflation the Universe was filled with a 'false vacuum', a higher energy state that, at least in our part of the Universe, decayed back to a ground state. The energy of this false vacuum drove the increased expansion. Alan Guth suggested that it might only decay in some parts of the Universe, creating 'bubbles' of different expansion rates. Each bubble would form its own isolated universe, one of many in a larger multiverse. In a similar vein, Andrei Linde developed a model of chaotic inflation that proposes an eternal multiverse arising from a quantum foam in which fluctuations can spark new Big Bangs and new periods of inflation in different regions. The vagaries of quantum fluctuations could, therefore, mean our Universe is just one among an infinite number of universes.

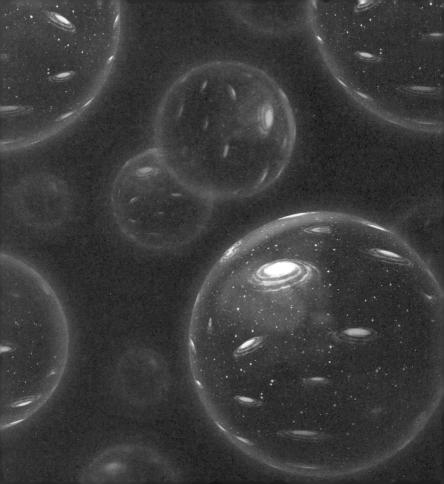

The expanding Universe

In the wake of the Big Bang and inflation, the Universe is still expanding today, a fact that was discovered by the American astronomer Edwin Hubble. Prior to 1925, nobody knew that there were galaxies beyond our Milky Way; most scientists assumed that the mysterious 'spiral nebulae' were part of our own star system. However, using what was then the world's largest telescope, Hubble resolved individual stars in these spiral nebulae. Using an ingenious method to calculate their intrinsic brightness he realized they were millions of light years away, and that spiral nebulae must be galaxies in their own right.

What was more, light from these remote galaxies was stretched to longer, redder wavelengths by expanding space and the Doppler effect (see opposite). Not only are galaxies generally moving away from Earth, but the more distant ones are moving away more rapidly, an effect known as Hubble's law. It turns out that the Universe is currently expanding at 22.4 kilometres (13.9 miles) per second per hundred million light years of space.

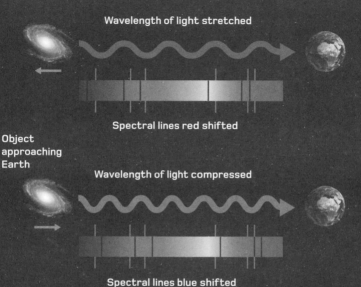

Object receding
from Earth

Wavelength of light stretched

Spectral lines red shifted

Object
approaching
Earth

Wavelength of light compressed

Spectral lines blue shifted

The Doppler effect alters the colour of light from receding or approaching objects, and is best detected when spectral lines are shifted out of place. On the largest scales, cosmic expandsion is also responsible for stretching and reddening the light of distant galaxies.

The accelerating Universe

The wide acceptance of the Big Bang theory meant that, even after the addition of inflation, most cosmologists assumed that the expansion of the Universe must have been gradually slowing down as its energy became more thinly spread. So it came as a startling surprise when, in 1988, two groups of astronomers found that the Universe wasn't slowing down at all, but was speeding up. The astronomers were probing the distant Universe for light from supernovae (exploding stars; see page 234), whose brightness can reveal their exact distance and therefore pin down the rate of cosmic expansion. Previous measurements of expansion had depended on objects in the relatively local Universe, but these more distant measurements showed how fast the Universe had been expanding over the billions of years since the light left these remote galaxies.

The accelerating rate of expansion shows that the Universe is gaining an energy boost from somewhere, and astronomers soon gave this mysterious accelerating factor a name: 'dark energy'.

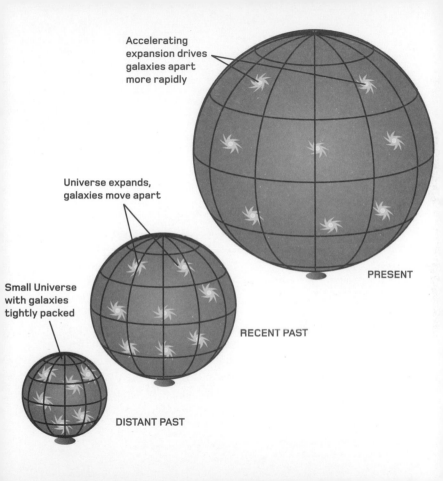

Accelerating expansion drives galaxies apart more rapidly

Universe expands, galaxies move apart

Small Universe with galaxies tightly packed

PRESENT

RECENT PAST

DISTANT PAST

Dark energy

What exactly is the force that is driving cosmic expansion to accelerate? The answer almost certainly lies in the fundamental nature of space on a quantum level. One of the leading contenders for the identity of this dark energy is a so-called 'cosmological constant' that describes a density of energy inherent to each point in space. As the Universe expands, there are naturally more points in space, and hence exponentially more energy to drive expansion. Unfortunately, calculations indicate that the most likely form of cosmological constant, called 'vacuum energy' (see page 140), is about 10^{-54} times too small to account for the effects of dark energy.

The other leading possibility is that dark energy is something called 'quintessence' which, if it exists, would be a quantum field that pervades the Universe and that can have varying strength in different regions of time and space. Quintessence could be attractive or, in the case of the accelerated expansion, repulsive. As yet, however, there is no independent evidence for its existence.

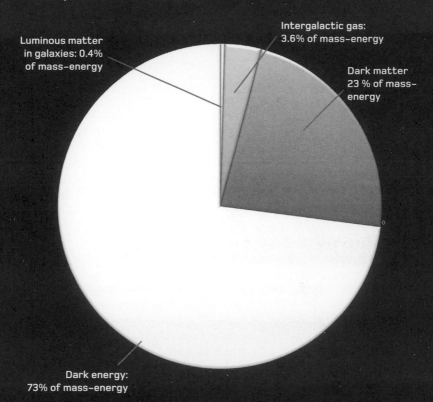

Composition of the Universe

Luminous matter in galaxies: 0.4% of mass-energy

Intergalactic gas: 3.6% of mass-energy

Dark matter 23 % of mass-energy

Dark energy: 73% of mass-energy

The death of stars

Astronomers seek out the stellar explosions of supernovae to measure the rate of the Universe's expansion because certain types explode with predictable levels of light. By measuring how bright they appear, astronomers can calculate their distance, and compare that with the rate at which they are moving across space. But supernovae are more than just distance markers. These violent bursts of destruction create extreme environments in which quantum effects reign supreme.

A supernova can manifest in two ways. The first is the explosion of a massive star that runs out of fuel (causing its core to collapse and a shockwave to blast its outer layers apart). The second is the violent collapse of a white dwarf (the remnants of a burnt-out, Sunlike star) into a much denser state. Both types of event can create a neutron star, within which atoms are broken apart into their subatomic components, the forces of electromagnetism are overcome and only the rules of quantum physics prevent a complete collapse.

Two paths to a supernova

Stellar explosion

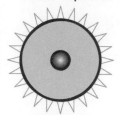

Massive, unstable star

Core exhausts fuel and collapses

Shockwave tears star apart

White dwarf collapse

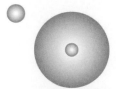

Sunlike star in binary system evolves into white dwarf

White dwarf pulls material away from companion star, increasing its mass

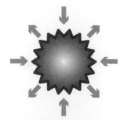

White dwarf becomes unstable and collapses into neutron star with burst of energy

Neutron stars

Stars generate energy through nuclear fusion reactions in their cores, governed by the laws of quantum physics. They begin life with a supply of hydrogen and helium, and temperatures in their core strip these elements of their electrons, creating an electrically charged plasma of atomic nuclei and free-floating electrons. Fusion steadily creates heavier elements by fusing the nuclei together, and in stars with eight or more times the Sun's mass, these reactions ultimately fill the star's core with iron, an element whose fusion absorbs more energy than it releases. Fusion abruptly halts, and with no energy to support it, the core collapses under its own gravity. Under normal circumstances, the collapse of a star's burnt-out core is halted by the Pauli exclusion principle, which creates a 'degeneracy pressure' between electrons and prevents them being compressed beyond a certain limit. However the collapse of a massive stellar core breaks these rules. As it dwindles to a diameter of just a few kilometres, extreme conditions break atomic nuclei into protons and neutrons and force electrons and protons together to create more neutrons. The resulting objects are called neutron stars.

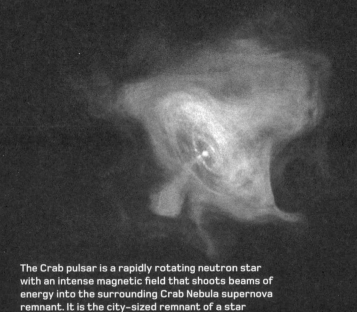

The Crab pulsar is a rapidly rotating neutron star with an intense magnetic field that shoots beams of energy into the surrounding Crab Nebula supernova remnant. It is the city-sized remnant of a star whose explosion was seen on Earth in 1054 CE.

Quark stars

Neutron stars are supported against gravity by a force known as neutron degeneracy pressure. This quantum property is determined by Pauli's exclusion principle, which on the level of neutrons states that no two particles with identical quantum states can share the same space. In a neutron star, all the lowest quantum states are filled. This creates a repulsive pressure, known as degeneracy, that prevents the neutrons from getting any closer together and halts the star's collapse.

However, if the force of the supernova or even the neutron star's own gravity is great enough, it can overcome the neutron degeneracy pressure. The continuing collapse overwhelms even the strong force, and breaks neutrons into their individual quarks, creating an object made out of exotic 'quark matter'. Such quark stars are still hypothetical, although some candidates have been identified. The confirmation of quark stars would provide a revolutionary new laboratory for testing models of particle and quantum physics.

Neutron star

Quark star

Core made of neutrons kept apart by neutron degeneracy pressure

Interior composed of quarks held up by quark degeneracy pressure

Black holes

Sometimes, the core of a star is so massive that, when it goes supernova, gravity can overwhelm neutron degeneracy pressure and even quark degeneracy pressure (hyopthetically created by strong-force interactions between quarks). The core of the star, many times more massive than the Sun, collapses to a single point of near-infinite density – a black hole. We call the centre of a black hole a 'singularity': the physics within it goes beyond our current understanding, but could be described by quantum gravity (see page 262). In accordance with Einstein's theory of general relativity, a black hole warps spacetime around it. Its gravity is so strong that even light straying too close cannot escape: hence, the singularity is surrounded by an invisible barrier called the 'event horizon'.

Black holes are not theoretical objects. Stellar-mass black holes from supernovae have been located in objects known as X-ray binary systems, while monstrous supermassive black holes, millions or billions of times the mass of the Sun, have been identified at the centre of many galaxies.

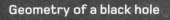

Geometry of a black hole

Ergosphere of distorted, rotating spacetime close to black hole

Flat spacetime at safe distance from black hole

Gravitational well

Event horizon where even light cannot escape from black hole

Singularity – collapsed point mass at centre of black hole

Hawking radiation

Although even light cannot escape the event horizon of a black hole, there's one caveat. In 1974, Stephen Hawking realized that virtual particles, forming on the inside edge of the event horizon thanks to the Heisenberg uncertainty principle, could potentially tunnel their way out. Hawking's idea was that when a particle-antiparticle pair is created, one falls into the black hole and gives the illusion of having negative energy, while the other tunnels its way to freedom. To conserve energy, it takes some from the black hole and becomes a real particle.

This phenomenon, as yet unobserved, is called Hawking radiation. Since every escaping particle carries with it some of a black hole's mass/energy, it means that the black hole is gradually evaporating. This causes something of a paradox. Does quantum information from objects that have fallen into the black hole just disappear from the Universe or is it conserved, and possibly even released in the Hawking radiation? Hawking himself is uncertain on the matter, but many other physicists believe that it is indeed conserved.

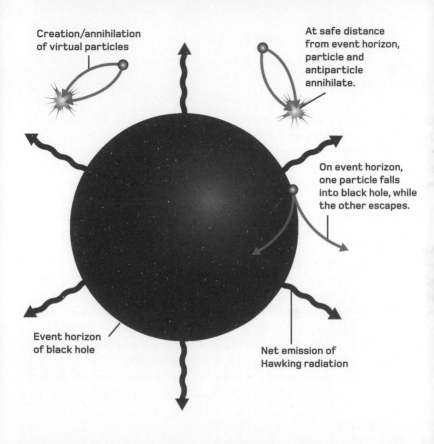

Creation/annihilation of virtual particles

At safe distance from event horizon, particle and antiparticle annihilate.

On event horizon, one particle falls into black hole, while the other escapes.

Event horizon of black hole

Net emission of Hawking radiation

Proton decay

So far as we know from observations, protons are eternal. Yet even though no one has ever seen a proton decay, does that mean they never do? Theories attempting to combine the fundamental forces of the quantum Universe into a 'grand unified theory' (GUT; see page 253) predict that protons decay with a half-life between 10^{34} and 10^{36} years. The Super-Kamiokande particle experiment in Japan has searched for proton decay and found nothing, constraining the half-life to at least 10^{35} years. Either way, it's an extremely long time and means proton decay, if it happens, is very rare.

This has consequences for the fate of the Universe. Matter itself will begin to disintegrate as protons decay into a neutral pion particle and a positron, or a neutral pion and a muon, which in turn will spontaneously decay into gamma-ray photons. If these theories are correct (and the Universe lasts that long), then more than one trillion trillion trillion years into the future all matter will have decayed into radiation.

Theoretical pathway
of proton decay

d
down quark

u
up quark

u
up quark

Neutral pion
particle

π^0

e^+
positron

γ —— Gamma rays —— γ

Vacuum decay

During the inflation epoch, a tiny fraction of a second after the Big Bang, the Universe is said to have been in a state of 'false vacuum'; that is, the quantum fields underlying the Universe were at a higher energy level, a bit like an electron in a higher energy level around an atomic nucleus. The ground state of the Universe is termed the 'true vacuum'. (In this case a vacuum means space almost devoid of energy, rather than air.)

The theory of eternal inflation suggests that some pockets of space are always in a false vacuum state, and there's circumstantial evidence that our Universe could be one of them. False vacuums are 'metastable': they can survive for a long time, but eventually decay, quantum tunnelling into a true vacuum. If this were to happen, it would create a bubble of true vacuum emanating from the point at which the tunnelling occurred, expanding at the speed of light and annihilating everything in its path until the entire Universe is a true vacuum. Fortunately, even if our Universe is in a false vacuum state, it's unlikely to decay for billions of years.

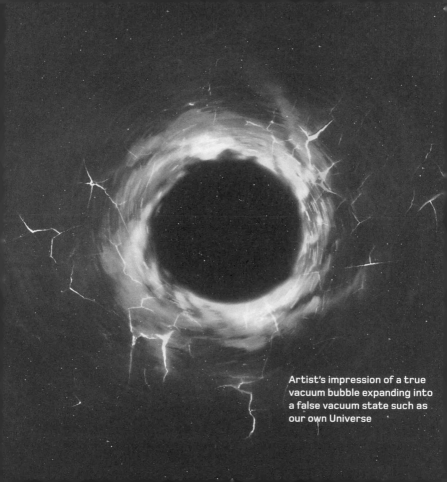

Artist's impression of a true
vacuum bubble expanding into
a false vacuum state such as
our own Universe

Fates of the Universe

Before the processes of vacuum decay or proton decay can destroy the Universe, dark energy and gravity may get the final say. If left unchecked, the cosmic expansion driven by dark energy (see page 232) could keep accelerating until space itself is torn apart in a 'Big Rip'. However, if scientists can show that dark energy has not remained constant in the past, this could indicate that, one day in the future, the rate of acceleration will decrease, sparing us from this fate.

Combatting dark energy in a cosmic tug-of-war is the force of gravity. Dark energy may not need to decrease by too much for the mass of the Universe to win out. Perhaps gravity will slow the expansion to a halt, creating a static Universe in which gas is spread so thinly that no new stars can form (a 'Big Chill' scenario). Alternatively, gravity might even reverse the expansion altogether, causing everything to come together again in a 'Big Crunch' governed by the unknown laws of quantum gravity (see page 262), and perhaps sparking a new Big Bang event in its aftermath.

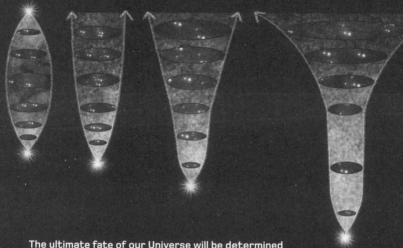

Collapse to a Big Crunch

Slowing expansion and Big Chill

Continued expansion with Big Chill

Accelerating expansion and Big Rip

The ultimate fate of our Universe will be determined by the contest between gravity and dark energy. Recent evidence that dark energy has grown stronger over the past few billion years suggests that a Big Chill, or perhaps even a Big Rip, is the most likely outcome.

Before the Big Bang?

Many cosmological models that utilize quantum mechanics suggest that the Big Bang may not have been the beginning of everything after all. Eternal inflation, for example, describes how a multitude of Big Bangs could keep happening in different parts of the Universe as a result of quantum fluctuations in the false vacuum (see page 246).

Another way for something to have existed before the Big Bang is if the Universe is cyclical. This would only happen if dark energy diminishes and there is sufficient matter for gravity to pull the Universe back to a 'Big Crunch'. During such an event, all the matter and energy in the Universe would be condensed down to a singularity of incredibly high temperature and density, in which the force of quantum gravity would once again reign supreme and perhaps begin the chain of events all over again. However, measurements of the mass density of the Universe so far seem to suggest that it is not above the required 'critical density', so we may be doomed to the cold fate of an expanding cosmos.

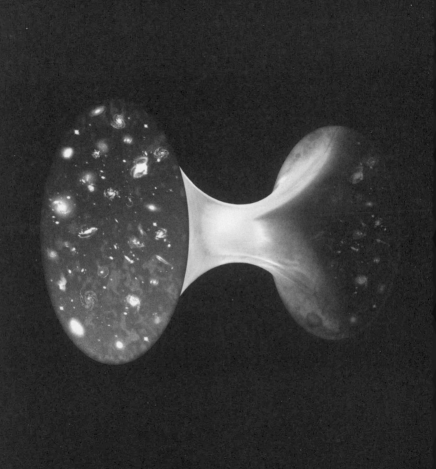

The theory of everything

Quantum mechanics unifies three of the four fundamental forces in nature: the strong and weak forces at work inside atoms, and the electromagnetic force. Gravity is excluded, and theoreticians investigating the operation of gravity on a quantum scale see it as the key to a grand 'theory of everything'. Such a theory would not only explain where the fundamental forces are coming from, but also the structure of the Universe and even, perhaps, its origins.

Unifying Einstein's general theory of relativity with quantum mechanics is not easy, however. Gravity operates on the largest scales, where the more mass there is, the more gravity there is. In contrast, quantum mechanics describes the physics of the very small — things like particles and individual photons of light. Quantum mechanics and gravity only come together in extreme environments — during the Big Bang or inside a black hole, for example. But if physicists can figure out how to unify them, the new physics revealed could prove revolutionary.

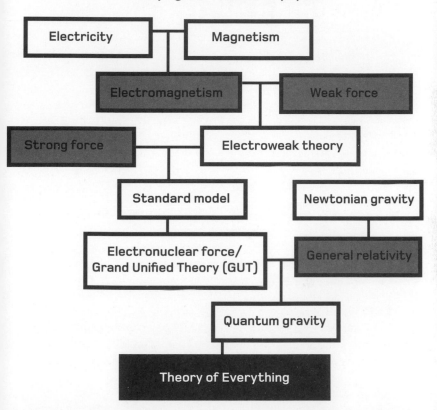

Unifying the theories of physics

Quantum field theory

Quantum field theory (QFT) is the foundation upon which the modern understanding of quantum physics is built. A 'field' in this context is a way of defining how a chosen quantity operates at every possible point in spacetime. For example, the electromagnetic field defines the electromagnetic force at any point in space and time. A quantum field is just the quantization of a classical field like electromagnetism.

QFT describes how fields are able to transmit their influence using force-carrying messenger particles called gauge bosons. Particles are simply afterthoughts to the field, a way of making real the field's quantization. A field theory is also handy when dealing with the unspecified number of virtual particles that result from Heisenberg's uncertainty principle when a particle has a relativistic energy, since the Schrödinger wave equation only works for one particle at a time. It's also worth noting that, since gravity is also a field, any theory that combines quantum mechanics with gravity is itself going to be a field theory.

Symmetry

Nature is full of symmetries and quantum mechanics is no different. When physicists refer to 'symmetry', they are discussing properties that are unchanged when they undergo a transformation. An everyday example of symmetry is a mirror image: hold some writing up to a mirror and some of the letters are reversed; the reflection has changed them, so they are not symmetrical. Other letters do not change and look exactly the same in the reflection. These are symmetrical.

In quantum field theory, 'gauge symmetry' is a special type of symmetry that explains why particles of a specific type, be they protons, electrons or quarks, are identical, or symmetric, to every other particle of their type. Gauge symmetry also plays a role in unifying the fundamental forces of nature: at increasingly high energies, forces become symmetrical and operate identically. First the electromagnetic and weak forces, then the strong force, and finally gravity all start to behave in the same way.

Common types of symmetry

Reflection

Rotation

Translation

Glide-reflection

Quantum electrodynamics

Q uantum electrodynamics (QED) is the field theory describing how the electromagnetic force interacts with matter. A common situation involves two electrons colliding and being repelled and scattered by their like charges. The force between them is a quantized electromagnetic field, carried by photons. To explain the theory visually, US physicist Richard Feynman developed Feynman diagrams (see page 202), essentially a pictorial depiction of the equation governing the interaction. Either side of the interaction in a diagram should balance in terms of mass/energy, charge, momentum and any other conserved properties.

The middle part of the diagram, where the interaction occurs, must incorporate all possibilities, like the multiple peaks of a wave function. However, Feynman diagrams offer a way to determine which process is most likely. The points where photons are emitted or absorbed are called vertices, and the more vertices there are, the less likely that particular process will happen.

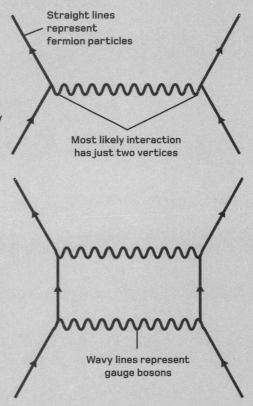

Feynman diagrams for the simplest (top) and a more complex version of an interaction between two electrons. The simplest interaction is also the most likely to occur.

Straight lines represent fermion particles

Most likely interaction has just two vertices

Wavy lines represent gauge bosons

Quantum chromodynamics

Quantum chromodynamics (QCD) explains the strong force that holds protons and neutrons together. In essence, it is the quantum theory of quarks, which exist within protons and neutrons in trios. This seemingly presents a problem, since two of each trio are either both up or both down quarks, with identical spin and charge. This violates Pauli's exclusion principle (see page 76), since two particles with the exact same quantum numbers should not be in the same place at the same time.

Physicist Murray Gell-Mann proposed that there must be another unknown quantum number at work to differentiate between the two up or two down quarks. He called this property 'colour', hence the term 'chromodynamics'. QCD results in two important properties, namely 'asymptotic freedom', which describes the strange effect of the strong force becoming stronger with distance rather than weaker, and 'confinement' which prevents particles with the colour property from existing individually.

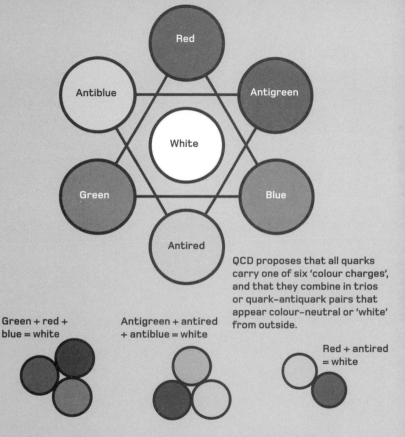

Red

Antiblue

Antigreen

White

Green

Blue

Antired

QCD proposes that all quarks carry one of six 'colour charges', and that they combine in trios or quark-antiquark pairs that appear colour-neutral or 'white' from outside.

Green + red + blue = white

Antigreen + antired + antiblue = white

Red + antired = white

Quantum gravity

Gravity is the only force not yet reconciled with quantum mechanics. Einstein's general theory of relativity doesn't describe it as a force, but as a distortion of spacetime by the mass of objects. Yet a theory describing it as a quantized field is required to explain how it operates on very small scales, in dense regions such as black holes or during the Big Bang.

One major problem is that, because Heisenberg's uncertainty principle prevents us from precisely pinning down both the position and momentum of particles, we cannot know the exact gravitational force they will feel. Another is that the bending of spacetime by relativity robs us of a fixed coordinate system: how can we determine the probability of a particle existing in a given location if space itself is constantly changing? If gravity can be quantized, then it should have a force-carrying boson particle of its own, but this hypothetical particle, the graviton, remains undetected, perhaps because its energy lies far beyond even that achievable in the Large Hadron Collider.

Quantum gravity may hold the key to understanding the mysteries at the heart of black holes and other extreme cosmic phenomena.

Electroweak theory

Above temperatures of a thousand trillion degrees, the electromagnetic force that governs light and the weak force that controls the radioactivity of elements become unified and symmetric. This 'electroweak' force, discovered by Steven Weinberg, Abdus Salam and Sheldon Glashow in the 1960s, existed in nature shortly after the Big Bang.

At distances of just 10^{-18} metres (a billion billionths of a metre), the weak force and electromagnetic force still retain much the same strength. At large distances, however, the weak force rapidly declines. This is because its force carriers, the W and Z bosons, are among the most massive particles known, so their 'virtual particles' do not travel far, while the photon, as carrier of the electromagnetic force, has no mass at all and hence a theoretically limitless range. Physicists suspect the radical difference in mass between force carriers is due to some kind of interaction with the Higgs field (see page 120), when symmetry was broken and the two forces separated.

An electroweak interaction recorded by the
CMS detector at the Large Hadron Collider

Loop quantum gravity

One possible solution to the problem of quantum gravity is that space itself is quantized. Physicists Lee Smolin and Theodore Jacobson proposed this idea in 1986, envisaging space as being formed by myriad interlinking quantum 'loops', each no bigger than the smallest size that is theoretically measurable (the so-called 'Planck length' of 1.6×10^{-35} metres). If such loops exist, then it means that space is granular at quantum scales. Networks of interlinked loops are known as 'spin networks', while a 'spin foam' describes how a spin network changes over time as a result of varying gravitational fields.

One advantage of this 'loop quantum gravity' (LQG) theory is that it removes the need to worry about precise location. Loops can be moved around by the warping of space without changing how they respond to gravity. However, LQG is still very much a work in progress, and it also makes no predictions regarding the graviton, a particle considered essential to a quantum field theory of gravity.

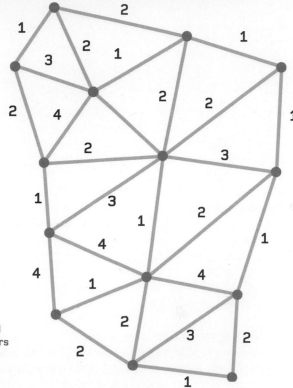

Spin networks are a type of diagram used to model interactions between particles and fields, and used by Smolin and others in the development of LQG.

String theory

A more popular rival to loop quantum gravity, string theory posits that all matter is made of tiny vibrating strings no longer than the Planck length. These strings vibrate at different frequencies, creating a variety of 'notes', to use a musical analogy. Each note manifests the quantum numbers of a different particle, giving rise to all the particles in the Standard Model. The theory depicts interactions between particles in terms of strings splitting apart and rejoining. Furthermore, it is dependent on the existence of the graviton and the maths will not work without it. Add the fact that the maths of string theory does not give rise to awkward mathematical infinities, and it makes for an extremely attractive theory of everything.

However, although some discoveries have been argued to corroborate it, string theory is untestable. Its maths may describe reality accurately, but as yet there is no way to experimentally confirm whether particles really are vibrating strings.

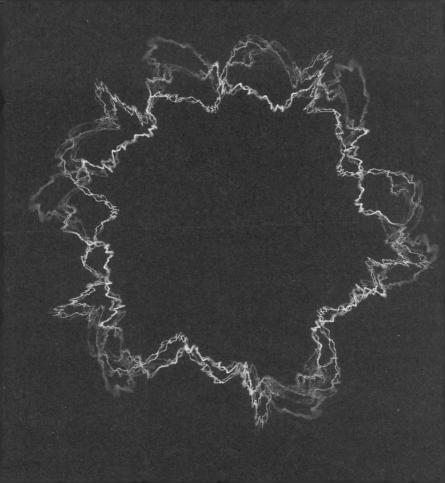

Planck epoch

What makes physicists believe that all four fundamental forces of nature really can be unified in a single quantum theory? The main reason is because scientists know that in the distant past they were united in just such a way. This moment, known as the Planck epoch, happened in the first instant after the Big Bang and before inflation, when the Universe was no more than 10^{-43} seconds old.

At the time, the Universe was still compressed into a minuscule volume, yet it contained the same amount of energy that it does today, leading to incredibly high temperatures of around 10^{34}°C. Under these circumstances, electromagnetism and the strong and weak forces, closely united with each other in a so-called 'Grand Unified Theory', merged togther with quantum gravity to act as a single superforce with uniform properties. The splitting apart of this superforce gave rise to every law of nature that we see in the Universe today.

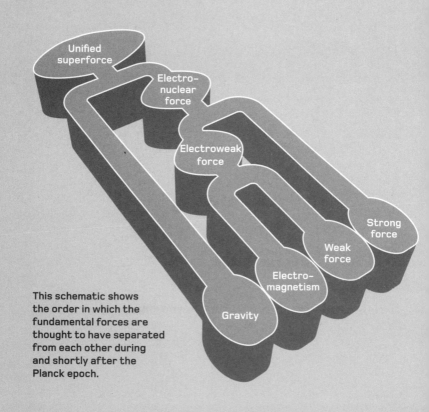

This schematic shows the order in which the fundamental forces are thought to have separated from each other during and shortly after the Planck epoch.

Symmetry breaking

At the end of the Planck epoch, 10^{-43} seconds after the Big Bang, the four elements of the primordial superforce began to separate, no longer acting as one. The forces split apart, one at a time, as temperatures dropped. Each separation marked a 'phase change' in the laws of physics, analogous to the changes we see when steam condenses into water and then freezes into ice. Those familiar changes involve a pause in the drop in temperature as reconfiguration of bonds releases energy, and something similar happened in these more fundamental transitions. The period of phase changes is known as symmetry breaking (the symmetry being the way the unified forces initially displayed identical strength).

The first force to break off was gravity, and energy released in the phase change created the quantum foam of spacetime. At 10^{-36} seconds, the strong force separated, releasing a burst of energy that may have driven cosmic inflation. Finally, electromagnetism and the weak force split at between 10^{-12} and 10^{-6} seconds as the Universe cooled past 10 quadrillion °C.

Phase transitions involve changes in the arrangement of matter. Though triggered by changes in temperature, they typically involve a brief pause in that change, as when ice melts into water at freezing point.

Supersymmetry

A hypothetical model of particle physics, called supersymmetry, says that the Standard Model is symmetric, with each boson symmetrically linked to a fermion, known as its 'superpartner'. The masses, charges and other quantum numbers of symmetric particles are identical; only their spin differs, since that is what defines bosons and fermions.

So why don't we see these superpartners in nature? Their apparent absence means that if supersymmetry is real, it must be a 'broken symmetry', such that the superpartners have far higher mass-energies between 100 and 1000 billion eV (higher than even the Higgs Boson). So why do physicists persist with this theory? Supersymmetry turns out to have many benefits essential to producing working models of string theory, and also offers a potential identity for the Universe's dark matter. What's more, it could also be a key player in unifying the four fundamental forces, since it causes their very different strengths to converge when traced back to the Planck epoch.

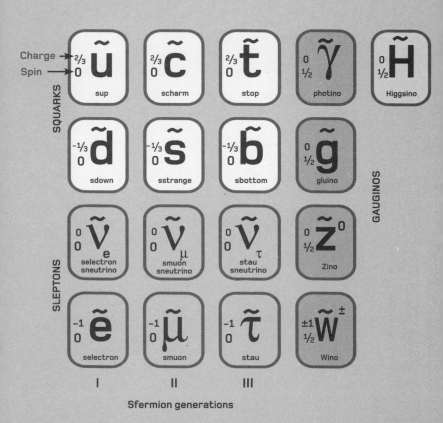

Superpartners of the Standard Model particles

Higher dimensions

We experience our Universe in four dimensions: three of space and one of time. So when the original equations of string theory came up with results that needed 26 dimensions of spacetime in order to work, there were naturally reservations. Things got a little better with the introduction of supersymmetry to string theory, which reduced the required dimensions to ten (nine of space, plus time). A recently developed model, called M-theory, unifies five rival versions of string theory, but needs 11 dimensions to work.

If any of these theories are correct, then where are the extra dimensions, and why don't we experience them? One possibility, known as compactification, involves them being wrapped up very tightly on such microscopic scales that we cannot detect them. Another option is that the dimensions are very large and that our three-dimensional Universe resides inside them as a sort of 'membrane' floating through higher-dimensional space.

Higher dimensions may be imperceptible
to us because they are compactified on
very tiny scales. An analogy is the way
we can curl a sheet of paper into a tube,
which appears as a one-dimensional line
when seen from a great distance.

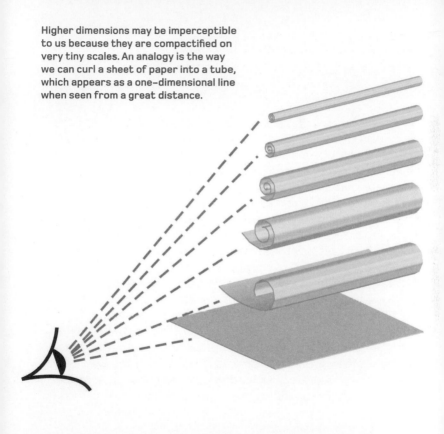

Calabi-Yau space

If string theory is correct, and the extra dimensions it requires compactified, where do they lurk? The answer may lie in a place called a Calabi-Yau manifold, named after mathematical theorists Eugenio Calabi and Shing-Tung Yau. The idea is that the entrance into six-dimensional space within the Calabi-Yau manifold is tiny, on the scale of 10^{-15} metres across. 'Unravelling' the manifold to make the effect of any of its dimensions measurable requires huge amounts of energy, but scientists at the LHC hope to find some evidence of Calabi-Yau space in energetic particle collisions.

Calabi-Yau space is appealing to proponents of superstrings, because the compactification of dimensions allows for a partially unbroken supersymmetry. String theory allows for many different types of six-dimensional spaces, and predicts that Calabi-Yau space should contains 'holes', corresponding to the number of particle families. If this is correct, we can narrow the possible solutions down to Calabi-Yau spaces with three holes (corresponding to fermions, quarks and bosons).

This graphical representation shows how a multidimensional Calabi-Yau space might be perceived in three-dimensional space.

Brane theory

A spin-off from string theory, brane theory describes a scenario involving a higher-dimensional space with extended dimensions, sometimes called hyperspace or 'the bulk'. A brane (derived from the word membrane) is a physical representation of a dimension, or collection of dimensions, in hyperspace. Individual objects are described by p-branes, where p is the number of dimensions involved. A point particle like an electron, with no physical size, would therefore be a 0-brane, a string would be a 1-brane as it exists in one dimension, and so on. Strings can either be looped or open-ended, and in the latter case the ends of the strings are attached to so-called D-branes, which are multidimensional objects moving through hyperspace.

According to the so-called 'braneworld' cosmology, our Universe is just such a brane. It has even been suggested that the Big Bang occurred when two branes collided, with the subsequent expansion of the Universe caused by the two branes then moving apart.

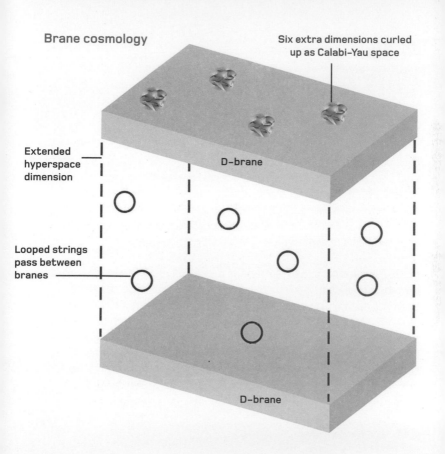

Brane cosmology

Six extra dimensions curled up as Calabi-Yau space

Extended hyperspace dimension

D-brane

Looped strings pass between branes

D-brane

AdS/CFT correspondence

We live in a Universe dominated by dark energy, a mysterious force accelerating cosmic expansion. A popular explanation is that dark energy is the cosmological constant, a hypothetical energy field that first cropped up in solutions to Einstein's field equations of spacetime. A Universe dominated by a positive cosmological constant is called a 'de Sitter space', after Dutch scientist Willem de Sitter. Anti de Sitter (AdS) space, in contrast, would have a negative cosmological constant causing expansion to decelerate.

In 1997, Argentinian Juan Maldacena made an astonishing discovery: if we extend an AdS into five dimensions, then 'our' four-dimensional Universe appears on its 'boundary surface'. Furthermore, there is a relationship between five-dimensonal gravity and a group of quantum field theories called conformal field theories (CFTs) in four dimensions. Maldacena's discovery marked a major advance in the search for a theory of everything, since it appears to confirm the holographic principle, a proposed property of quantum gravity that 'encodes' its higher-dimensional properties onto four-dimensional spacetime.

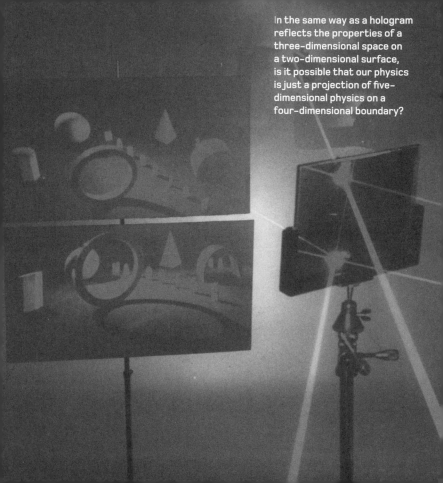

In the same way as a hologram reflects the properties of a three-dimensional space on a two-dimensional surface, is it possible that our physics is just a projection of five-dimensional physics on a four-dimensional boundary?

The best theory?

For several decades, string theory has been the leading contender for a theory of everything that unifies the fundamental forces, explains the Standard Model and describes a quantum theory of gravity. It does have detractors, however, with its unproven extra dimensions a particular focus for criticism. The discovery of superpartner particles (see page 274) would provide good, albeit indirect, evidence for string theory. These particles should have energies in the range of 100 to 1,000 billion electronvolts. Presently, the LHC can only probe the bottom end of this energy scale, however, and so far there's no sign of even the lightest proposed superpartner.

This might seem like good news for loop quantum gravity, in which supersymmetry is optional. But LQG has its own problems; critics point out that spin networks do not incorporate time, and also fail to explain the Standard Model. It seems that there's still a long way to go to find a theory of everything.

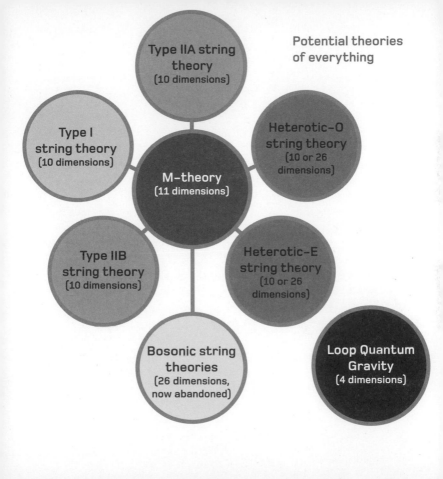

Potential theories of everything

Type IIA string theory
(10 dimensions)

Heterotic-O string theory
(10 or 26 dimensions)

Type I string theory
(10 dimensions)

M-theory
(11 dimensions)

Type IIB string theory
(10 dimensions)

Heterotic-E string theory
(10 or 26 dimensions)

Bosonic string theories
(26 dimensions, now abandoned)

Loop Quantum Gravity
(4 dimensions)

Many-worlds interpretation

One of the most stunning consequences of quantum theory is that our reality might not be the only one; there could be an infinite number of parallel universes in which every possible event can occur. According to the Copenhagen interpretation (see page 54), the wave function is simply our best attempt at describing the inherent uncertainty in quantum mechanics. Once an observation is made, it collapses to produce a single outcome. In 1957, however, physicist Hugh Everett III suggested an astonishing alternative: what if the wave function never collapses and, instead, reality itself diverges, allowing every possibility it describes to happen somewhere in an alternative universe? Everett's idea, now known as the 'many-worlds' interpretation, is a leading alternative to Copenhagen.

It's not the only quantum theory of parallel worlds. Eternal inflation, sparked by quantum fluctuations, could create myriad new universes, each with their own characteristics and realities and possibly even an infinite number of you.

Types of multiverse

The multiverse is the name given to a collection of different universes that are possibly infinite in number. Max Tegmark, a cosmologist at the Massachusetts Institute of Technology (MIT), hypothesizes that there are four different types (see opposite). The simplest, Level 1 multiverse, relies simply on the overall Universe being very, very big, much larger than the 'observable Universe' whose edges are limited by the speed of light (see page 222).

Factoring in cosmic expansion, our observable Universe is around 96 billion light years across. If we took every single atom within it, put them in a bag and shook them around, there is only a finite number of ways in which they could be rearranged. In an infinite Universe this means every possible arrangement will eventually be repeated – not just once, but an infinite number of times. If the idea is correct, then somewhere out there, at an unimaginable distance, there is another you at the centre of a sphere that looks just like our own observable Universe.

Tegmark's 4 levels of multiverse

1. The extension of normal spacetime beyond the limits of our observable Universe

2. The multiverse of Universes with different physical properties produced by processes such as eternal inflation

3. The multiverse of parallel Universes arising from the many-worlds interpretation of quantum mechanics

4. The 'ultimate ensemble', a set of mathematical structures capable of describing any possible multiverse including those in Levels 1 to 3

The inflationary multiverse

If a Level 1 multiverse has 'more space' and is similar to our own but repeated to infinity, Tegmark's Level 2 multiverse is more varied. Eternal inflation (see page 226) predicts that parts of the Universe are constantly 'budding off', thanks to quantum fluctuations that drive new bouts of inflation. This creates a multitude of new universes, expanding away so fast that nothing in our Universe could ever reach and enter them.

If string theory is broadly correct, then each of these new universes could potentially have completely different laws of physics to our own. This is because the equations of string theory may have some 10^{500} potential solutions, each of which could describe a universe with different laws of physics – a different mix of dimensions, different strengths of fundamental forces and different fundamental particles. Chaotic inflation could produce an infinite number of such universes and, furthermore, many of these could be Level 1 multiverses in their own right.

This artist's impression shows new universes budding off each other in an everlasting process of inflation.

The uncollapsible
wave function

In the early days of quantum physics, many scientists were unhappy with the Copenhagen interpretation, since certain readings of it implied that vast expanses of the cosmos could exist in probabilistic limbo until observed. Among them was Hugh Everett III, whose many-worlds interpretation suggests that wave functions don't really collapse when observed, they just present an *illusion* of collapse.

Everett pointed out that it's not only the object being observed that is in a state of quantum flux; so, too, is the observer. If an electron has a possibility of existing at one of several points, then the observer also has a wave function describing the possibility of their observing the electron in each location. The electron and the observer's quantum states are 'entangled', with the outcome of one related to the outcome of the other. Each possible outcome for the observer is superposed over the other, and in each outcome the observer sees their version of the wave function collapse. In general, the wave function is uncollapsible, but we can experience only one of its outcomes.

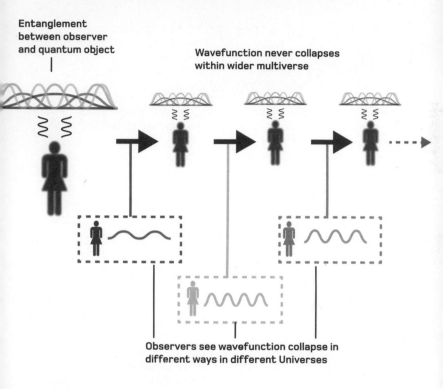

Entanglement between observer and quantum object

Wavefunction never collapses within wider multiverse

Observers see wavefunction collapse in different ways in different Universes

The many-worlds multiverse

According to the many-worlds interpretation (see page 286), every possible outcome of every wave function occurs somewhere – but where does this all take place? The answer, it seems, lies in the existence of parallel worlds, classified by Max Tegmark as the Level 3 multiverse.

Unfortunately the many-worlds theory has little to say on the subject of how the continual divergence into parallel universes happens, or where these infinite parallel realities exist relative to each other. Since our entire world is built on quantum foundations, everything that happens is probabilistic in nature, and every possibility branches off from each other, like branches on a tree. In one universe, Schrödinger's cat lives, in a multitude of others that branch off from the first universe every second, the cat dies. But the cat will only ever know the universe in which it survives. This gives rise to an intriguing 'test' of the many-worlds theory, known as the quantum suicide experiment.

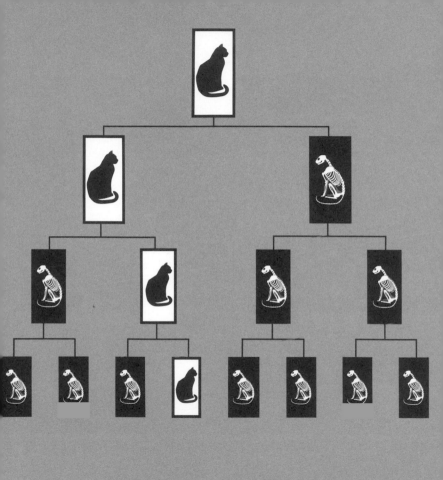

Quantum suicide

In the many-worlds interpretation, Schrödinger's cat never experiences a universe in which it dies. That's the astonishing conclusion of a thought experiment called quantum suicide. The experiment (not to be attempted!) is a kind of quantum Russian roulette, a modification of Schrödinger's original idea, but with the experimenter themselves standing in for the cat. A gun is connected to a particle in a state of quantum superposition and measured every second. If the particle is found to be in one particular state, the gun fires a bullet; if it is in the other state, the gun holds fire.

Ordinarily, the experimenter might be lucky and survive the first few times, but within a few seconds they will be shot and killed. However, in the many-worlds interpretation, the experimenter survives every time; the wave function never collapses from their point of view, so they always perceive a universe in which they survive. Only an outside observer will see the wave function collapse and the experimenter die, while the experimenter lives on in a parallel reality.

A testable theory?

The many-worlds interpretation provides a compelling solution to counterintuitive aspects of the Copenhagen interpretation. It removes the paradox of Schrödinger's cat, and does away with the necessity for the Universe to be observed in order to exist in a given state. It also offers an explanation for the 'fine-tuning' problem.

However, without observational evidence to support it, the many-worlds interpretation has received criticism for being an untestable theory. The scientific method on which all modern science relies is based on empirical observation and testable predictions, but no one has yet conceived an experiment that could test the multiverse theory because, so far as we know, the different universes would not interact after splitting. With no satisfactory explanation for exactly how different quantum universes branch off from one another, at present it seems that belief in many worlds depends on how much faith you put in the logic and mathematics behind the theory.

Cyclical universes

The question of what came before the Big Bang has often been dismissed as meaningless; the Big Bang created everything, including time, so there should have been nothing at all before it, not even empty space.

But this is not necessarily true. For a start, if eternal inflation (see page 226) might today be continually branching new universes off from our own, then presumably our Universe would once have budded off another, even older universe. Brane theory (see page 280) gives us another alternative, the cyclical universe. Cosmologists Neil Turok and Paul Steinhardt ask us to imagine two parallel branes moving towards one another, then colliding and rebounding. The collision causes a big bang and the dark energy that is causing our Universe to expand is a force felt between the two branes as they move away. Eventually, the branes move closer once more and our Universe begins to contract, resulting in a 'Big Crunch' that immediately creates a new Big Bang, and the cycle repeats again.

An artist's impression shows two D-brane Universes approaching one another in hyperspace (see page 280).

The anthropic principle

One of the strangest aspects of our Universe is the way in which fundamental constants that lie behind many fundamental physical processes seem peculiarly fine-tuned to create a cosmos capable of supporting life. For example, if the strong force was just slightly weaker, quarks would not be able to hold together and form baryons. If it was slightly stronger, it would have caused all the hydrogen in the early Universe to fuse into helium, robbing stars of their fuel supply. The speed of light, the charge of the electron and the strength of gravity also have values that are just right for life.

Cosmologists explain this fine-tuning with an idea called the anthropic principle. The 'weak' form of the principle argues that we should expect to measure values like these, since we could not exist in a Universe that is not suitable for life. The 'strong' form, in contrast, looks for a reason behind the fine-tuning: perhaps it is a consequence of the theory of everything, or perhaps our Universe is indeed one among an infinitely varied multiverse, not all of which have given rise to life.

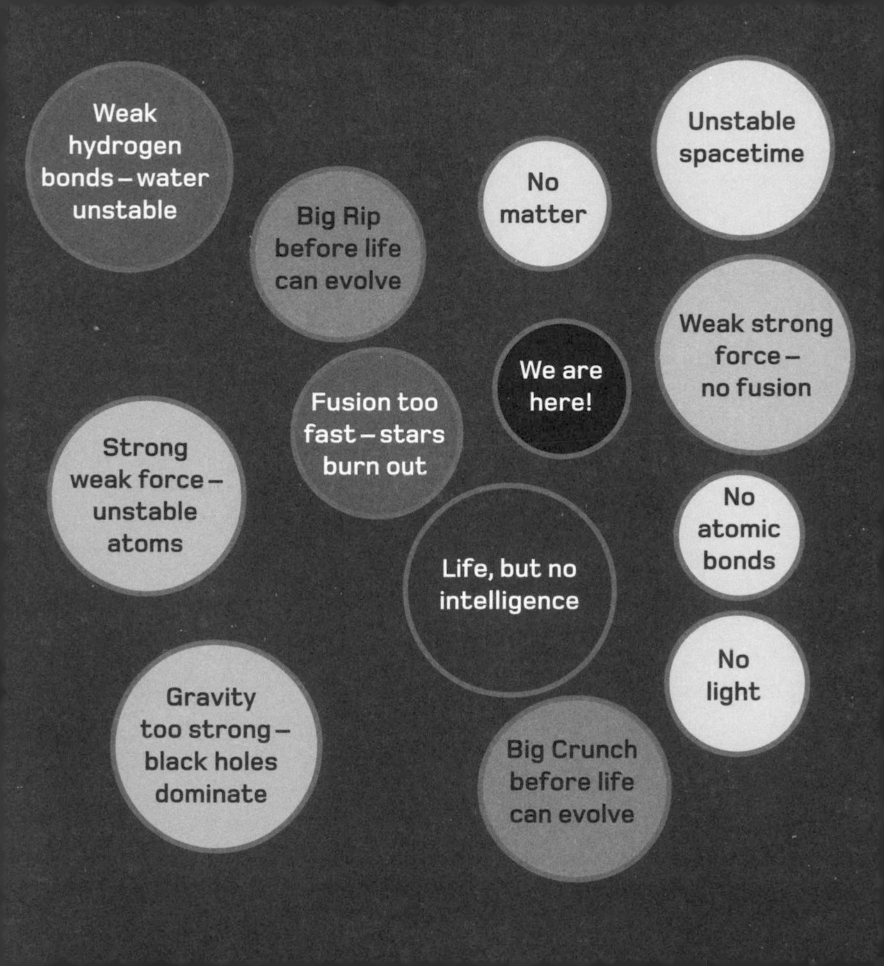

Weak hydrogen bonds – water unstable
Big Rip before life can evolve
No matter
Unstable spacetime
Weak strong force – no fusion
We are here!
Fusion too fast – stars burn out
Strong weak force – unstable atoms
Life, but no intelligence
No atomic bonds
No light
Gravity too strong – black holes dominate
Big Crunch before life can evolve

Playing dice

When Albert Einstein declared that 'God doesn't play dice with the world', he was bemoaning the apparent randomness of the Copenhagen interpretation's probabilistic wave function. This has consequences far beyond whether light or electrons are particles or waves; the Heisenberg uncertainty principle means that at the quantum level, nature is fundamentally random and cannot be predicted to any degree of accuracy.

Einstein utterly rejected this notion. To him, the apparent randomness just meant that our understanding of quantum physics was incomplete; there must be more information buried within the properties of particles to describe their behaviour in a predictable, deterministic manner. However, Einstein admitted his objection was based on gut instinct and our intuition is, of course, biased by our observations of an everyday world that is predictable and deterministic. Ultimately, Einstein was proved wrong about quantum mechanics by an experiment of his own devising.

Quantum entanglement and the EPR paradox

In 1935 Einstein, along with fellow physicists Boris Podolsky and Nathan Rosen, set out their concerns about the Copenhagen interpretation in what became known as the EPR paradox. Suppose an atomic nucleus decays into a pair of particles that move apart in opposite directions. Because they formed in a state of superposition, their quantum properties are entangled. So if, for example, one particle's spin is measured and found to be 'spin down', then according to Copenhagen the other particle's wave function must simultaneously collapse and force the other electron to be 'spin up', even if by now it is on the far side of the Universe.

Einstein famously described the phenomenon as 'spooky action at a distance', but since information cannot travel faster than light, he could not see how it was possible. Yet experiments have shown that entanglement is exactly what happens: quantum mechanics operates on a principle of 'nonlocality' that goes against our classical understanding of physics.

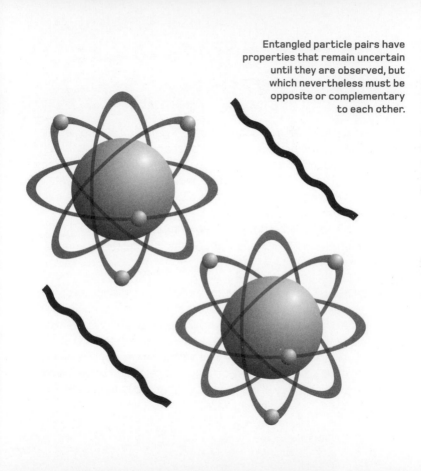

Entangled particle pairs have properties that remain uncertain until they are observed, but which nevertheless must be opposite or complementary to each other.

Hidden variables and Bell's theorem

Irish-born John Bell expanded the EPR paradox in a series of thought experiments from which he made predictions that have subsequently been borne out. Einstein, Podolsky and Rosen believed that there must be some 'hidden variables', as-yet-undetected properties, that carry the information to let each particle know which state it is in. If these factors existed, then both the faster-than-light communication paradox of entanglement and the uncertainty inherent in the Copenhagen interpretation could be avoided.

Bell put the EPR paradox through a strict mathematical test, now known as Bell's theorem. The nature of a particle's quantum spin means that the probability of measuring a given spin depends on the angle from which it is measured, so Bell performed a statistical analysis, calculating the odds of measuring a given spin from a given angle. He could find no evidence for a relationship between the probabilities and the angles that suggested hidden variables existed. Instead, entanglement must be real.

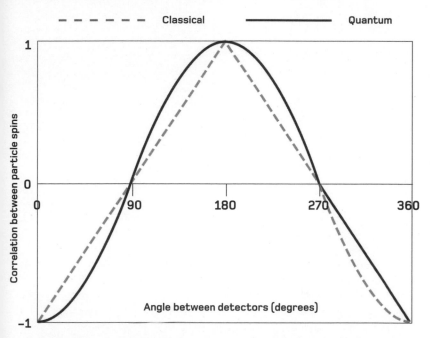

Experiments to test Bell's theorem by measuring the spin of entangled particles produce results that match the quantum mechanical, rather than classical, distribution. A classical distribution could be explained by hidden variables, but the true quantum distribution cannot – hence entanglement is real.

Defying causality

The main reason Einstein was wrong about entanglement is that he assumed that cause and effect operated on the basis of 'locality', where information propagates outwards from the location of the 'cause' at the speed of light. This is one of the most intuitive principles in physics: if you commit an action, the consequences of that action should naturally follow.

Quantum entanglement, however, seems to operate on a principle of nonlocality; the distance between entangled particles doesn't matter. It's hugely counterintuitive and still not fully understood, but it means that normal rules of cause and effect no longer apply. John Bell likened this bizarre behaviour to his friend Dr Reinhold Bertlmann, who liked to wear odd socks of different colours, one blue and one green, on randomly different feet each day. If one morning you saw him wearing a blue sock on his right foot then you could instantly know that the green sock was on his left foot without taking the time to look, just as entanglement can defy causality by conveying information in a non-local way.

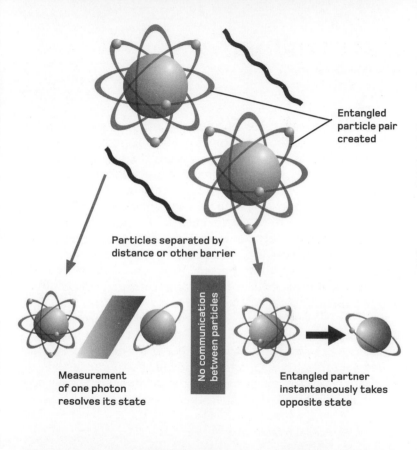

Entangled
particle pair
created

Particles separated by
distance or other barrier

No communication between particles

Measurement
of one photon
resolves its state

Entangled partner
instantaneously takes
opposite state

Determinism

Classical physics relies on the principle of determinism, the idea that the state of an object is completely determined by its earlier states. Take, for example, kicking a football: the ball's physical properties (such as its shape and weight) and the forces acting upon it (the strength of the kick, location of contact, wind strength and so on) all affect where the ball is eventually going to land. If you have access to all such information, the behaviour of the ball is perfectly predictable. However, if that ball exists in a series of quantum states, then the Heisenberg uncertainty principle means that its future states are inherently unpredictable and capable of changing instantly.

Entanglement is another means of determining the future state of a particle without that linear progression of cause and effect (see page 310). This is the fundamental difference between classical physics and quantum physics: one is deterministic, while the other is probabilistic.

Faster than light?

It's little wonder that Albert Einstein was not a fan of the notion of quantum entanglement, since his special theory of relativity declares that nothing can travel through the Universe faster than the speed of light. However, if information about entangled quantum states can travel faster than light (see page 310) does this mean that other information can also be communicated instantly across vast distances?

Einstein's postulate regarding the speed of light survives because of a technicality. It's not information regarding the quantum state of a particle that is being communicated faster than the speed of light. Instead, it is some kind of signal for the particle to *reveal* its quantum state that is propagating faster than the speed of light. The information is already contained within the particle's wave function. This subtle distinction means that there may be limits to how we can apply quantum entanglement to our advantage — we may still be prevented from sending 'useful' information at faster-than-light speeds.

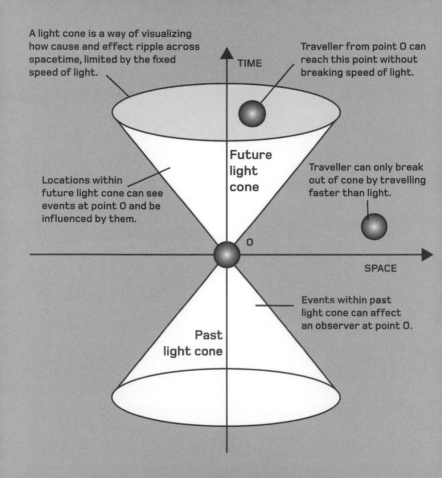

A light cone is a way of visualizing how cause and effect ripple across spacetime, limited by the fixed speed of light.

TIME

Traveller from point O can reach this point without breaking speed of light.

Future light cone

Locations within future light cone can see events at point O and be influenced by them.

Traveller can only break out of cone by travelling faster than light.

O

SPACE

Events within past light cone can affect an observer at point O.

Past light cone

Quantum teleportation

Quantum entanglement opens the door to teleportation of a kind. Physical objects cannot be sent instantaneously over great distances, but their quantum states can, allowing for the creation of replicas. For teleportation to work, we need three objects: two of them (particles X and Y) are entangled and begin to move apart. At some indeterminate distance from each other, X encounters particle Z. Quantum information from Z transfers to X, and X's quantum state is then instantly communicated from X to Y, transforming Y into a replica of Z.

One complication is that the quantum state of Z is destroyed in the process. This may make human teleportation, if it ever becomes a reality, a somewhat scary process. What's more, a *Star-Trek*-style transporter would need a supply of atoms at the destination ready to take on quantum information. The sheer amount of information involved in sending any large object would mean the process would take a very long time, and decoherence might create further stumbling blocks.

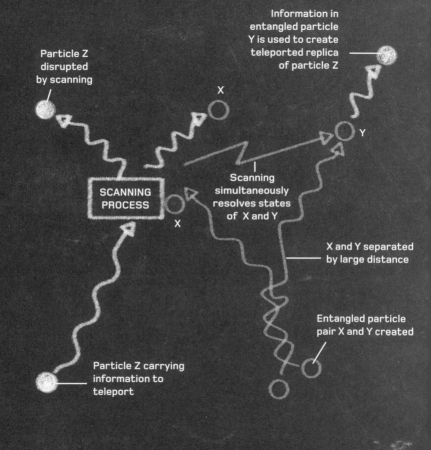

Information in entangled particle Y is used to create teleported replica of particle Z

Particle Z disrupted by scanning

X

SCANNING PROCESS

Scanning simultaneously resolves states of X and Y

Y

X

X and Y separated by large distance

Particle Z carrying information to teleport

Entangled particle pair X and Y created

Teleportation experiments

Quantum teleportation isn't just a theoretical fancy; scientists have already succeeded in teleporting particle information. The first successful experiment was conducted in 1998: just five years after the first theoretical thesis on the possibility had been written, researchers succeeded in teleporting the quantum state of a photon across a table-top. In 2004, scientists teleported an atom for the first time.

Since then, the range of teleportation has grown substantially. The current record for teleporting the quantum state of a photon stands at 144 kilometres (89 miles), achieved by a team led by Anton Zeilinger of the University of Vienna. That experiment was conducted across 'free space', but in 2015 American scientists at the National Institute of Standards and Technology were able to teleport the quantum states of photons down 102 kilometres (63 miles) of fibre optic cable. In the future, such techniques could prove useful for setting up secure communication systems using quantum entanglement.

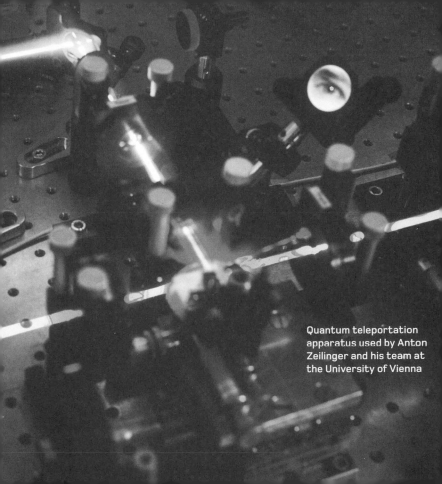

Quantum teleportation apparatus used by Anton Zeilinger and his team at the University of Vienna

Quantum time

In the everyday world of classical physics, the flow of time coincides with increasing entropy (the natural and inevitable loss of order in thermodynamic systems, see page 22). On quantum scales, however, entropy doesn't have quite the same hold that it does on macroscopic scales. Thanks to Heisenberg's uncertainty principle, a handful of particles could turn from an ordered, low-entropy state into a disordered, high-entropy state and back again almost at random. So what defines time on the quantum level?

Seth Lloyd of the Massachusetts Institute of Technology (MIT) believes that the flow of time is defined by an increasing loss of information. Decoherence and the collapse of wave functions are certainly irreversible ways of losing quantum information, but at its heart, says Lloyd, is entanglement. Imagine a cooling cup of tea. In Lloyd's picture of time, the tea's atoms gradually become entangled with their surroundings, moving the system towards greater equilibrium with the Universe, but losing the tea's quantum information in an irreversible process.

Time running backwards?

The principle of cause and effect seems locked to the forward flow of time, but aspects of the quantum world suggest things aren't always that simple. For example, in the dual-slit experiment, interference fringes are caused by wavelike behaviour (opposite, top). If you were to measure which slit each photon passes through (observing the photons as particles and not waves), the interference fringes would disappear (opposite, bottom). Suppose, however, that you change the experiment to measure which slit entangled photons pass through only *after* they have passed through it. Cause and effect says that you should be able to see the interference fringes, yet that's not what happens. Instead, we still observe the photons acting as particles: it seems that somehow the measurement in the present has affected the particle's behaviour in the past. Is this phenomenon, known as retrocausality, evidence for information travelling backwards in time? Perhaps, but most physicists believe it to be a result of quantum effects rather than actual time travel.

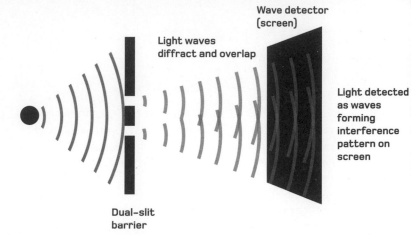

Wave detector (screen)

Light waves diffract and overlap

Light detected as waves forming interference pattern on screen

Dual-slit barrier

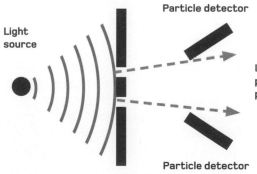

Particle detector

Light source

Light detected as photons on straight-line paths through slits

Particle detector

Boltzmann brains

Although we describe the flow of time as linked to irreversible processes, in truth, nothing is irreversible given enough time. The Heisenberg uncertainty principle means that there's a tiny chance of such processes reversing themselves; for example, if two canisters containing two different gases are mixed, there is a tiny probability, given aeons of time, that all the atoms will unmix and end up back in their respective canisters.

Another bizarre consequence of the passage of aeons are 'Boltzmann brains', first put forward by 19th-century physicist Ludwig Boltzmann. Boltzmann believed that we live in a chance fluctuation of low entropy and relative organization in a high-entropy Universe, and that other low-entropy fluctuations could naturally lead to the appearance of consciousness. Although Boltzmann had no knowledge of the quantum realm, there's a quantum mechanical analogy to this in the form of the quantum fluctuations that fill space. Given enough time, such fluctuations could fashion anything, even a conscious entity.

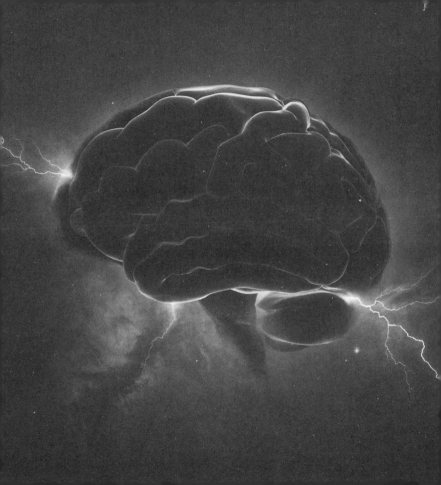

Quantum mechanical applications

Although the bizarre behaviour of particles in the quantum world seems remote from everyday experience, this does not mean that quantum physics is an abstract field that is of interest to theoretical physicists alone. In fact, nothing could be further from the truth: quantum physics is a practical science that has integrated itself into numerous aspects of our day-to-day lives. It lurks everywhere, from our electronics to our telecommunications, and from our smart phones to mundane visits to the supermarket.

Some technologies deliberately take advantage of quantum mechanical effects, while others were invented and applied long before the theory behind them was fully understood. Without quantum mechanics, much of the technology we take for granted in the modern world would not have come to exist. Quantum science also operates in living things, underying many chemical processes vital to life. Perhaps it even acts as the basis for our consciousness.

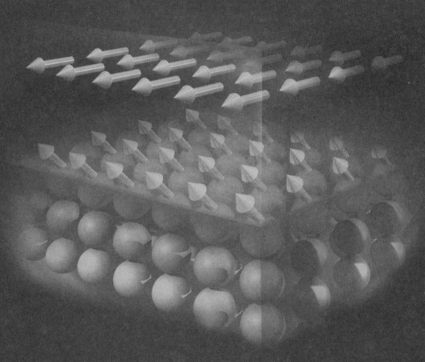

Harnessing quantum behaviour on a large scale allows scientists to produce revolutionary materials such as 'topological insulators' that allow current to flow through the surface, while remaining impervious to magnetic fields.

Lasers

Ubiquitous in modern technology, lasers are powerful beams of light that owe their unique properties to the fact that their individual photons are all 'coherent'. This means that the peaks and troughs of their waves are locked in step with each other, allowing them to form intense, tightly focused beams.

The word laser is an acronym for 'light amplification by stimulated emission of radiation'. It relies on a material called a lasing medium, in the form of a crystal or a gas. When electrons in the medium's atoms are energized by an electric field or intense light, they jump to a higher energy level. Normally, they would emit photons of identical wavelengths at random as they naturally dropped back to the ground state. However, here the surrounding laser 'cavity' traps these photons, bouncing them back and forth through the medium. As the photons interact with electrons each time, they trigger stimulated emission, forcing an electron to release another photon with identical properties to the first and amplifying the overall beam.

How a laser works

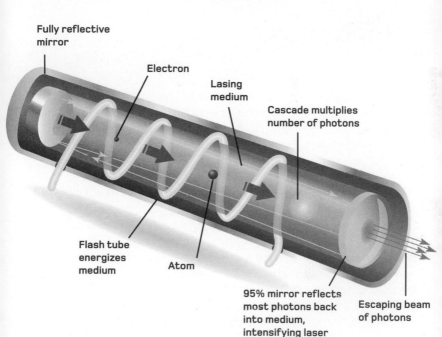

Fully reflective mirror

Electron

Lasing medium

Cascade multiplies number of photons

Flash tube energizes medium

Atom

95% mirror reflects most photons back into medium, intensifying laser

Escaping beam of photons

Scanning tunnelling microscope

The wavelike properties of electrons can be used to image objects on much smaller scales than is possible with visible light (see page 40). The scanning tunnelling microscope takes this a step further, making use of quantum tunnelling. It plays a key role in medical research and microchip manufacturing, among numerous other applications.

The microscope is essentially a stylus with an extremely fine tip that ends in a single atom that is brought within an atom's-width distance of the sample. A voltage is then applied that excites electrons in the surface, causing some of them to quantum tunnel across the distance between the surface and the tip and generating a so-called 'tunnelling current'. The tip then scans across the surface, moving up and down with respect to its contours to ensure that the tunnelling current remains constant. By monitoring the up-and-down movements of the stylus, the microscope builds up an image of the surface at the atomic scale.

Principle of the scanning tunnelling microscope

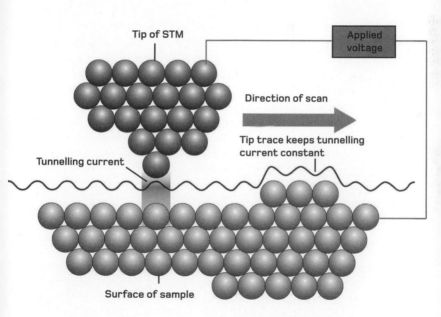

Magnetic resonance imaging

If you've ever been to hospital for a magnetic resonance imaging (MRI) scan, then you've experienced applied quantum physics at work in your very own body. MRI takes advantage of the fact that protons within hydrogen atoms inside your body's water and fat have a quantized spin that 'points' in one of two directions, each with a slightly different energy (see page 108). During an MRI scan you move through a short cylinder that applies a strong magnetic field. The field aligns most of the proton spins in the direction of the field, while those with higher spin energy are aligned in the opposite direction.

A rapidly varying radio-frequency magnetic field is then applied, which the lower-spin energy protons absorb, causing their spin to 'flip' to the higher state. When the magnetic field is turned off, the protons return to their lower state and emit radio waves that are detected by the scanner. Protons in different tissues return to their lower states at different rates, allowing doctors to differentiate between organs and monitor their health.

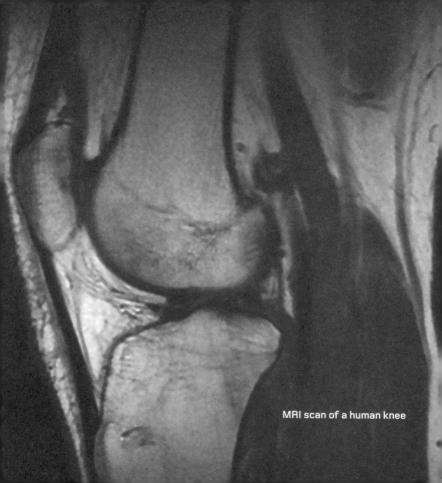
MRI scan of a human knee

Electronics

Modern microelectronics are built on silicon chips and the electric currents that move around them. Harnessing tiny flows of charged electrons through relatively small collections of atoms, they're a tangible example of quantum physics in action.

The electrons in silicon, as in any solid object, are distributed in quantized energy bands that dictate how that solid object conducts electrical current. The structure of the bands is unique to each material. By 'doping' silicon with small amounts of other elements, engineers can alter its conducting properties to suit a variety of applications, creating semiconductor materials that will only allow electricity to flow in certain directions or under certain conditions. Layered semiconductors can be used to build diodes, transistors and other electronic components that are mere nanometres across, yet are capable of performing simple 'logical' functions. Placed alongside each other on silicon chips, these components can be fashioned into the complex integrated circuits that are the basis of most modern technology.

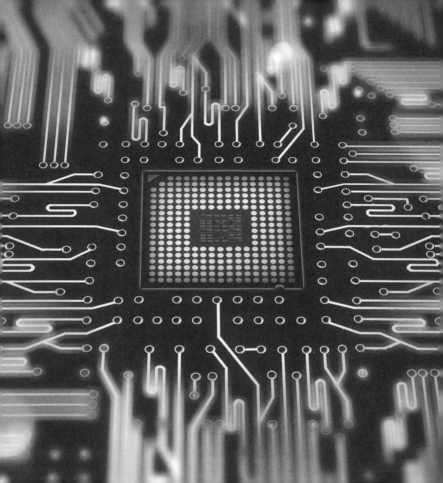

Flash drives

The humble computer memory stick is another quantum mechanical marvel. Its 'flash-drive' memory stores bits of digital data (1s and 0s), using a device known as a floating gate transistor. This contains two distinct logic-gate circuits – a 'control' gate that governs the flow of current through the transistor (like an on/off switch) and a floating gate that acts as a memory cell. To preserve its state, the floating gate is electrically insulated from the rest of the transistor by two thin oxide layers.

When we save a bit of data onto a memory stick, our computer sends a signal that applies a strong voltage across the transistor. This causes electrons to quantum tunnel their way across the oxide layers into the floating gate. Here, they become trapped and the data becomes stored in the insulated memory cell. To delete the data, a voltage is applied in the other direction so that the electrons can tunnel their way back through the oxide layer.

Structure of a floating gate transistor

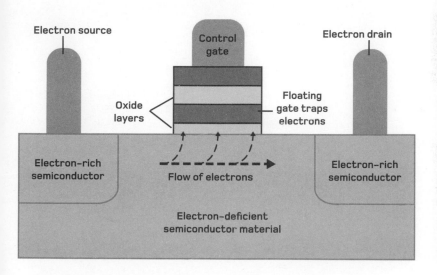

LEDs

Light emitting diodes (LEDs) are another ubiquitous feature of everyday life that operates on quantum principles. Inside an LED bulb is a semiconductor chip that only conducts electricity in certain conditions. The LED semiconductor is made of two layers of a crystalline material, such as gallium arsenide or gallium nitrate, mixed with other elements that slightly alter its conducting properties. This mixing leaves one of the layers with an excess of high-energy electrons and the other with many spaces at lower energies for the electrons to fill.

Between the two layers is a gap known as a *p-n* junction (*p* refers to the layer with spaces, *n* refers to the layer with the excess electrons). The *p-n* junction is a diode, meaning the electrons can only flow one way when a voltage is applied. As they cross the *p-n* junction, the electrons have to shed quanta of energy in the form of light, causing the diode to illuminate. The wider the junction, the greater the quantum jump, and the higher the energy and shorter the wavelength of light emitted.

Structure of an LED

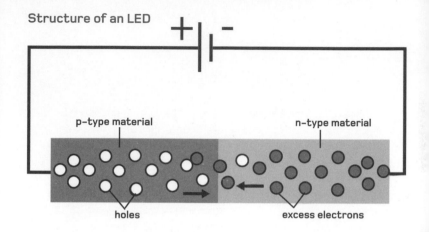

p-type material

n-type material

holes

excess electrons

Energy levels at the p-n junction

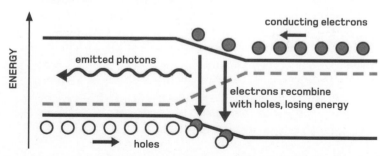

ENERGY

conducting electrons

emitted photons

electrons recombine
with holes, losing energy

holes

Atomic clocks

The most accurate timekeepers in the Universe, atomic clocks are unsurprisingly reliant on quantum principles. Magnetic fields and blue laser light are used to cool individual atoms of beryllium, caesium or strontium to extremely low temperatures at which the atoms are hardly moving. A red laser is then shone onto the atoms, its wavelength specifically attuned to the amount of energy required for their electrons to make a quantum jump to a higher energy level.

Once the electrons have absorbed a photon and jumped, they immediately emit a microwave photon and drop back down. As long as the atoms remain illuminated by the red laser, the electrons keep jumping up and down within a precise period, like a pendulum ticking off the seconds. Pulses of emitted microwaves create a measurable signal with an accuracy of one lost second every 300 million years. Even more accurate are 'quantum clocks' that measure the vibrational states of cooled ions of beryllium or aluminium. These clocks only lose a second every 3.86 billion years.

GPS satellites such as the European Galileo system carry atomic clocks for precise timekeeping.

Quantum cryptography

Secure encryption of data plays an increasingly important role in our information age. It usually involves algorithms based on long strings of numbers that are virtually impossible to break using traditional computing techniques. In the future, however, quantum computers may easily crack these codes, so systems are likely to rely on undecipherable quantum encryption.

In a typical quantum cryptography system, the sender encodes photons with binary data by adjusting their spin alignment. The receiver passes the photons through a filter (either + or x-shaped), and openly asks the sender if they chose the correct filter at each stage (see opposite). This tells the receiver the correct filter sequence, but the non-quantum information relayed is meaningless to eavesdroppers without context. Any attempt to intercept the signal will alter the spin states of the photons, showing that someone has been interfering. For even greater security, photons don't need to be beamed at all: the information could, instead, be sent using quantum teleportation.

Alice puts photons through filters to encode information before sending them to Bob. Photons cannot be intercepted without disrupting the message.

Bob puts photons through one of two filters and measures result.

Bob openly sends Alice a list of the filters he used.

Alice confirms which filters were right or wrong.

| YES | YES | NO | NO | YES | YES |

| 1 | 0 | 0 | 1 | 1 | 0 |
| | | (Not 1) | (Not 0) | | |

Bob can now deduce the filters Alice used,
and find the original information.

Telecommunications

Modern telecommunications are built around microwaves, lasers and optical fibres. Lasers in particular are a quantum phenomenon created by manipulating the way in which photons are emitted by electrons as they jump between energy levels (see page 328). Their intensity allows digital signals, in the form of pulses of laser light, to be transmitted across huge distances by fibre optic cables without losing strength.

In the near future, with data security becoming ever more important, tamper-proof quantum cryptography is sure to be more widely implemented acros telecommunications networks. In 2016, China launched the first quantum communications satellite, named Mozi. The satellite incorporates quantum key encryption and, if successful, will create an unhackable wireless network. The initial steps in setting up quantum-encrypted links to and from a distant satellite are complex, but in the next few years the Chinese government intend to have the first quantum communications network running between Europe and Asia.

Radiometric dating

An ingenious application of the quantum phenomenon of radioactivity (see page 128), radiometric dating uses the probabilistic decay of radioactive atoms to determine the age of everything from rocks to organic matter. The best-known method is carbon dating, widely used by archaeologists. A radioactive form of carbon, carbon-14, is continually produced in Earth's atmosphere as particles from space collide with nitrogen atoms.

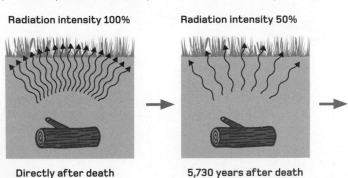

Radiation intensity 100%

Radiation intensity 50%

Directly after death

5,730 years after death

All living things contain a small amount of radioactive carbon that is constantly being recycled in and out of the environment.

Once an organism dies, however, this exchange stops, and the quantum process of radioactive decay takes hold. Carbon-14 has a half-life of 5,730 years, meaning it takes that long for half of the atoms in a sample to decay. Scientists can therefore measure the surviving amount of carbon-14 and work backwards to find the age of a sample. Carbon-14's half-life is relatively short, limiting the technique to relics dating back just 50,000 years, but similar techniques can be used to date billion-year-old rock samples.

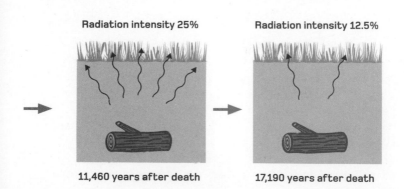

Radiation intensity 25%

Radiation intensity 12.5%

11,460 years after death

17,190 years after death

Quantum dots

Quantum dots are tiny pieces of semiconductor (usually silicon or germanium), just a few dozen atoms across. Atoms in the dot are so close that their electrons influence each other. However, because Pauli's exclusion principle forbids them to share the same quantum states, a new arrangement forms, creating new energy levels around the dot, rather like the electron orbitals around a single atom. For this reason, quantum dots are sometimes referred to as 'artificial atoms'.

As in an individual atom, electrons in the dot can absorb photons, jumping to higher energy levels and then emitting a photon as they drop back down, which causes the dots to glow. The size of the dot dictates the colour it glows: in larger dots the energy levels are more closely spaced, so the energy of the photon is lower and the light redder. Smaller dots have more broadly spaced energy levels and so produce higher energy, bluer photons. Quantum dots can be used as biosensors, in solar cells, or even as LEDs in next-generation television sets.

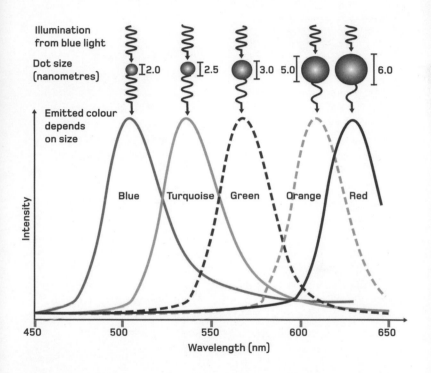

Quantum dots: size and colour

Illumination from blue light

Dot size (nanometres)

2.0   2.5   3.0   5.0   6.0

Emitted colour depends on size

Intensity

Blue   Turquoise   Green   Orange   Red

450   500   550   600   650

Wavelength (nm)

Superfluids

When certain liquids, such as liquid helium, are cooled to just a few degrees above absolute zero, they lose all frictional resistance. Given a little momentum, they will keep flowing uphill or will creep out of containers and over obstacles. Set a superfluid spinning and it creates quantum vortices that carry quantized angular momentum, and can keep swirling indefinitely.

Superfluids are Bose-Einstein condensates (see page 114), systems in which atomic bosons drop to the lowest possible energy level and hence avoid collisions, dramatically lowering their viscosity. As quantum solvents, superfluids can dissolve chemicals into clumps of just a few molecules, surrounded by a 'quantum solvation shell' that allows them to rotate freely. This proves useful for studying individual gas molecules. Frictionless superfluids have also been used in a high-precision gyroscope, and also as a means of 'trapping' electromagnetic radiation: interaction between photons and superfluids can slow the speed of light to just 17 metres (56 ft) per second.

FERMIONS

BOSONS

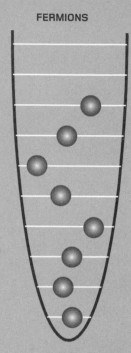

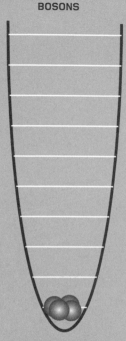

Fermions obey Pauli's exclusion principle – particles are forced to occupy different quantum states.

Cooled bosons can all fall into lowest-energy state, allowing them to exhibit identical behaviour such as superfluid properties.

Superconductivity

When some metals, such as lead, niobium, mercury and rhodium, are chilled to a few degrees above absolute zero, they experience a sudden drop in electrical resistance to practically zero. They become superconductors, capable of holding an electrical current without losing energy, in theory for billions of years. Superconductors also repel magnetic fields, which is the secret behind magnetic levitation.

Inside the metals, ions (charged atomic nuclei) are arranged in a lattice structure, surrounded by electrons. Normally, ions vibrate and collide with the electrons flowing past them, creating electrical resistance. But when cooled below a critical temperature, the electrons begin to form pairs that defy Pauli's exclusion principle by having similar quantum states. The energy of these 'Cooper pairs' dramatically lowers and, in the lattice structure, an energy gap opens up above the electrons. Because they don't have enough energy to cross the gap, they can't collide with the ions, so there is no electrical resistance.

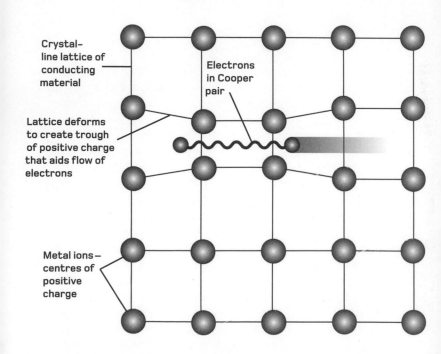

Inside a superconductor

Crystal-line lattice of conducting material

Electrons in Cooper pair

Lattice deforms to create trough of positive charge that aids flow of electrons

Metal ions – centres of positive charge

Quantum chemistry

Since it deals with molecules rather than individual atoms or particles, chemistry is generally considered to be a step up in scale from particle physics. Nevertheless, the quantum properties of atoms still have an effect on many aspects of chemistry. Chemical bonds are formed by the exchange or sharing of electrons between different atoms in order for each to achieve a relatively stable shell configuration (either full or half full), so it's understandable that quantum physics, by changing our understanding of the electrons and their orbits, also has an impact on our understanding of these bonds.

German physicists Walter Heitler and Fritz London used the newly minted Schrödinger wave equation to model the structure of the bond between two hydrogen atoms as early as 1927. Today's quantum chemists use a variety of different techniques, including computer modelling, to better understand how electron properties are distributed around more complex molecules, and how this can affect their larger-scale properties.

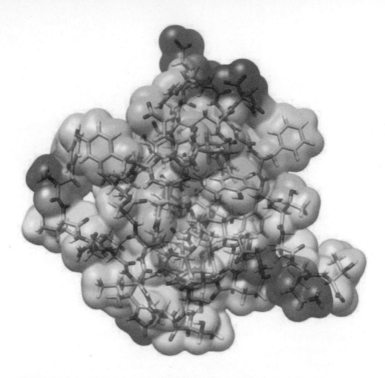

Using quantum mechanics to model the distribution of electrons, scientists can better understand the shape and functionality of complex chemicals such as this insulin molecule.

Quantum biology

All living things rely on the transfer and conversion of energy to keep themselves alive, and wherever energy is being converted, a quantum process is usually involved. The growing field of quantum biology seeks to explain biological processes through the prism of quantum mechanics.

Most functions of plant and animal life rely on chemical reactions that are themselves predicated on the quantum behaviour of electrons. Absorption of light can be used to generate chemical energy or send information to sensory organs. Neurons in our brains, meanwhile, are the nodes of chemical and electrical networks that operate on a quantum level. Enzymes, the biological workhorses that catalyse chemical reactions in our body, seem to use quantum tunnelling to help move electrons through otherwise insurmountable energy barriers around molecules. If we can understand how these processes work, we may be able to create artificial catalysts to generate energy and form new molecules in environmentally friendly ways.

On a fundamental level, biological functions such as the sending of nerve signals between synapses depend on chemical and electrical processes that are rooted in quantum mechanics.

Biological compasses

Some animals display a sixth sense that makes them aware of Earth's magnetic field. This is most obvious in migrating birds, which follow field lines to navigate across the planet.

For birds, one possible explanation is the presence in the sensory system of an iron-oxide mineral called magnetite, which is highly magnetic, allowing individual grains to align with Earth's own magnetism. Another explanation, however, invokes quantum mechanics. Certain proteins, sensitive to blue light, create a pair of 'radicals', highly reactive atoms or molecules with a single valence electron, rather than a pair. The effect of Earth's magnetism on their quantum spins causes the blue-sensitive proteins to remain active for longer and creates a colour shift in vision that a migrating bird can detect. It has even been suggested that valence electrons in each of the radicals could be entangled, ensuring that when one aligns with the magnetic field, so does the other.

Quantum photosynthesis

Perhaps surprisingly, one of the key processes that permits life on Earth also owes its success to quantum physics. Plants gain energy through photosynthesis, using energy from the Sun to convert water and carbon dioxide into glucose. The key to this process is the green pigment molecule chlorophyll and, specifically, elements called chromophores that capture the energy of sunlight in the first place.

When a photon is absorbed by chlorophyll, its energy creates molecular vibrations in a pair of chromophores that can only be described in quantum terms. The vibrations transport the energy around a leaf's cellular structure, and the efficiency of this energy transport is increased when the energy of a pair of vibrating chromophores matches their vibrational transitions, leading to the exchange of a quanta of energy. Remarkably, chlorophyll demonstrates this quantum behaviour in warm temperatures where we might expect it to be drowned out by other molecular vibrations. There's much we can learn from the humble tree leaf.

Quantum vision

Our eyes are biological sensors for detecting photons of light, so it's unsurprising that the vision process involves quantum physics. The retina at the back of the eye is lined with photoreceptor cells that convert photons of light into electrical signals using a chemical called retinal, which changes its structure when it absorbs light energy. This is the first step on an electrochemical pathway that ends with signals to the brain. However, in a biological analogue of the photoelectric effect, a photon must carry a certain quantized amount of energy in order to stimulate the retinal.

This explains a biological oddity. Our warm bodies produce large amounts of thermal (infrared) radiation, which inevitably leaks into our eyes. A million times more photons enter our eyes from our bodies than from the outside world, so why do we not see all this thermal radiation when we close our eyes? The answer is that, even though there are more of them, none of these thermal photons carry sufficient energy to stimulate retinal.

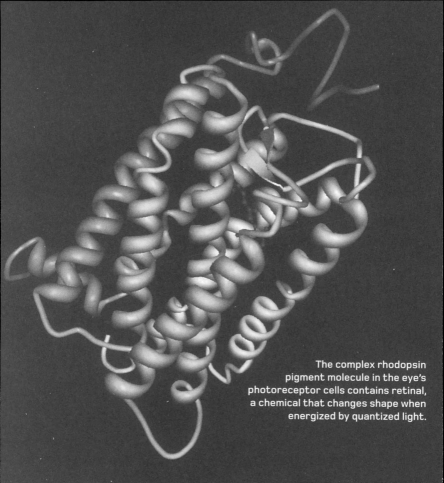

The complex rhodopsin pigment molecule in the eye's photoreceptor cells contains retinal, a chemical that changes shape when energized by quantized light.

Quantum consciousness

Could quantum physics lie at the root of human consciousness? Numerous physicists have speculated on these lines, including Niels Bohr and Eugene Wigner. The current champion for quantum consciousness is British physicist Roger Penrose. Together with an anaesthesiologist named Stuart Hameroff, Penrose has proposed a theory, called orchestrated objective reduction, that describes consciousness as a consequence of quantum gravity.

Penrose and Hameroff's idea is that quantum gravity manifests itself as spacetime vibrations inside tiny protein polymers called microtubules that reside in the neurons of the brain. A superposition of quantum states generated by the microtubules decays steadily rather than instantly, creating the moment-to-moment awareness of consciousness. In 2014, Penrose and Hameroff went further, claiming the rhythm of brain waves as evidence for the presence of spacetime vibrations within the microtubules.

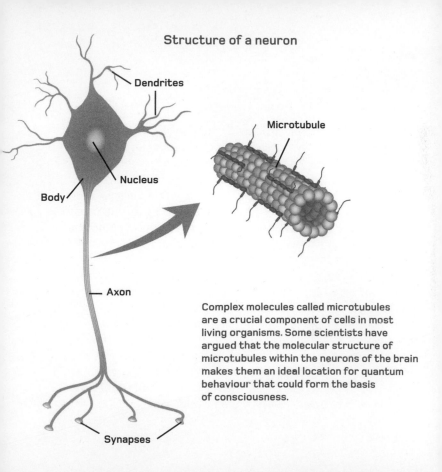

Structure of a neuron

Dendrites

Microtubule

Nucleus

Body

Axon

Synapses

Complex molecules called microtubules are a crucial component of cells in most living organisms. Some scientists have argued that the molecular structure of microtubules within the neurons of the brain makes them an ideal location for quantum behaviour that could form the basis of consciousness.

Against quantum consciousness

Many scientists have poured cold water on claims that human consciousness arises from quantum effects. Chief among these sceptics has been Max Tegmark of the Massachusetts Institute of Technology (MIT). Pointing out that the brain is a hot and very complex structure, he ran calculations suggesting any quantum superpositions that might arise in the brain would decohere faster than neurons can signal each other. This means that, should these quantum states exist, they could have no effect on brain processing.

Since Tegmark's analysis, however, studies have shown that living creatures can indeed use quantum effects to their benefit. These include photosynthesis in plants and the magnetoreception sense of migrating birds. Ultimately, the human brain is far too complex for us to model properly yet, which means there is still space for theories of quantum consciousness alongside the better-supported theories that our brain can be described by classical physics.

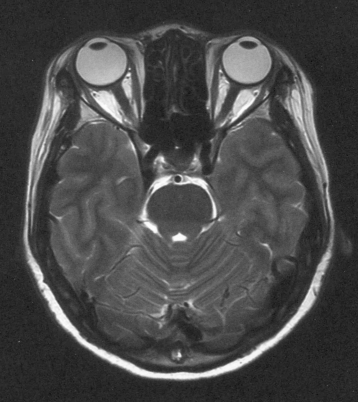

Is the human brain simply too complex and noisy for tiny quantum effects to make their influence felt?

No free will?

From a philosophical point of view, the most significant effect of quantum mechanics is whether or not it allows us free will. Many quantum physicists and philosophers believe that determinism, which says the Universe is predictable, results in all our actions and decisions being essentially predictable, too (if, of course, one had the computing power, a complete understanding of the processes involved and access to all necessary information about the Universe). Conversely, some also suggest that the probabilistic randomness of quantum mechanics effectively removes free will: if nothing can be predicted with accuracy, we don't really get any say in what happens as a result of our actions.

In either case, free will would be an illusion, and German physicist Sabine Hossenfelder has proposed the existence of 'free-will functions', hidden laws that could give rise to something that appears to be free will. Whether that distinction is enough to affect our own perception, however, is another matter.

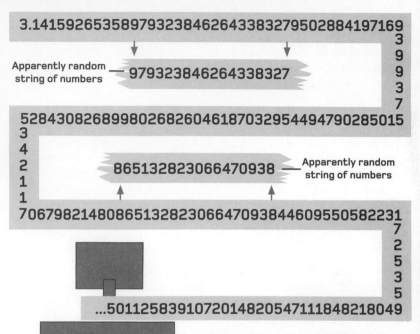

3.14159265358979323846264338327950288419716 9
3
9
9
3
7

Apparently random string of numbers — 979323846264338327

52843082689980268260461870329544947902850153
4
2
1
1

865132823066470938 — Apparently random string of numbers

70679821480865132823066470938446095505822317
2
5
3
5

...501125839107201482054711184821804 9

Printer

Sabine Hossenfelder compares her 'free will functions' to a machine that sequentially prints one digit of pi every second: if you read part of the printout from such a machine without knowing anything about it, the digits would appear random and unpredictable.

Quantum computing

Quantum computers promise to change the world in ways we can't imagine. Our information age sees us swamped in data – from social media to the results of scientific experiments – and traditional digital computers struggle when analysing huge amounts of information. Quantum computers, however, have the parallel processing power to take on these challenges.

While digital computers store information as binary 'bits' that take a value of either 0 or 1, quantum computers use the superposition of quantum states of particles to store information in elements called qubits. Superposition boosts the processing speed of quantum computers: while an ordinary computer works on just one computation at once, a quantum computer can be working on millions of computations simultaneously. In the future, these powerful devices will be able to sift and analyse enormous amounts of data, solving complex mathematical problems that can be applied to areas such as modelling the environment, curing disease and investigating the quantum world itself.

Qubits

Think of a qubit as a quantum of information – the simplest and smallest unit of information possible. The difference between a qubit and a digital bit is that, whereas a normal bit can be in only one of two states (0 or 1, true or false, yes or no), a qubit can exist as both 0 and 1, true and false, yes and no. This is because its quantum states are superposed, like Schrödinger's cat, until a measurement is made. Qubits can be individual atoms, ions, electrons, Bose-Einstein condensates, superconducting circuits called Josephson junctions or even photons of light.

The information in a qubit is encoded into its quantum properties, such as the spin of an electron or the polarization of a photon. The number of possible states equals 2^N for N qubits, so two qubits can process four states simultaneously, and six qubits can process 64 states simultaneously. Each qubit can ultimately produce just one answer when measured, but the superposition of states provides extraordinary processing power.

Classical and quantum bits

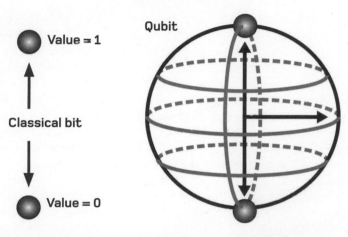

Value = 1

Classical bit

Value = 0

Qubit

A normal binary digit of information can only take on two states, 1 or 0.

A quantum qubit can exist in a superposition of 1 and 0 states, allowing an array of qubits to process huge amounts of information simultaneously.

Types of quantum computer

Physicists developing quantum computers don't expect to build a top-of-the-range model immediately. Instead, the development of quantum computers is expected to pass through three stages or milestones. The most basic, called a quantum annealer, considers variable quantum states as something like a topographical map with hills and valleys (see opposite). While devices capable of this step have been built, the technique is only useful for specific problems, and quantum annealers have not proved to be significantly faster than ordinary computers.

The next step, a so-called 'analog quantum computer', would be faster than a regular computer. Such a machine would operate with just 50–100 qubits and, again, could only solve a few types of problems. But it would be an important milestone on the way to a true universal quantum computer. Equipped with around 100,000 qubits, such a device would be exponentially faster than normal computers.

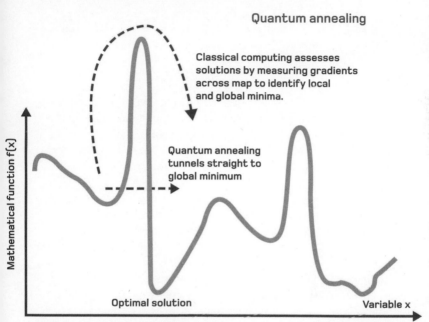

Quantum annealing considers mathematical problems in a way analogous to an elevation map, with hills and valleys, and a solution at the bottom of the lowest valley on the map (the 'global minimum'). A classical computer would have to search the entire map to find the right solution, but a quantum computer can effectively 'tunnel' through the hills to find the answer in moments.

Problems of decoherence

The biggest problem facing quantum computers is decoherence, the decay of the wave function when a qubit's quantum state is measured (see page 176). Decoherence would eradicate a qubit's superposition so that, instead, of being in a state of both 0 and 1, it would be forced to take on one value or the other.

Decoherence turns a quantum computer into a regular classical computer, and it will be hard to avoid. Qubits will have to be kept isolated from outside interference that would cause their wave functions to decay. Entanglement offers one possible way of measuring the state of the computer without disturbing the qubits doing the processing. However, many scientists developing quantum computer systems take the attitude that decoherence is something to manage; it's always going to be present, and there's a certain amount that can be tolerated. This can be done by having a computer with a large number of qubits, so that the error rate caused by decoherence is small compared with the number of qubits.

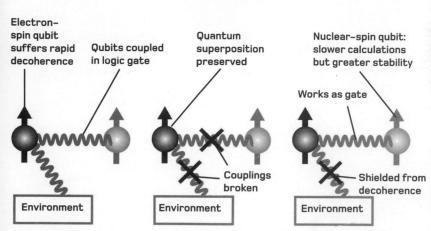

| Decohering logic gate | Decoupling protection | Protected gate |

Decohering logic gate

Decoupling protection

Protected gate

Electron-spin qubit suffers rapid decoherence

Qubits coupled in logic gate

Quantum superposition preserved

Couplings broken

Nuclear-spin qubit: slower calculations but greater stability

Works as gate

Shielded from decoherence

Environment

Environment

Environment

A two qubit logic gate (see page 380) loses fidelity through interactions with its environment.

Isolating qubits preserves information, but prevents them from functioning in a logic gate.

'Protected' gates can entangle two different qubits together while isolating the system from decoherence.

Controlling qubits

Isolating a quantum computer's qubits in order to avoid decoherence requires some means of trapping and holding them without causing their wave functions to decay. Computers that use atoms as their qubits can use a grid of lasers called an optical lattice to create potential wells where the beams intersect, trapping the atoms in these regions. Electrically charged ions, meanwhile, could be confined by electromagnetic fields, and might convey information through their collective motions as their charges influence each other.

Quantum dots (see page 348) can be used to control the electrons that arrange themselves in orbits around them, but quantum computers based on light are more problematic, since photons don't interact with one another. Mirrors and devices called beam-splitters might be one way of confining the light, as are so-called 'Rydberg atoms' – large atoms that can collectively slow light to a crawl – paving the way for quantum computer 'circuits' made from light itself.

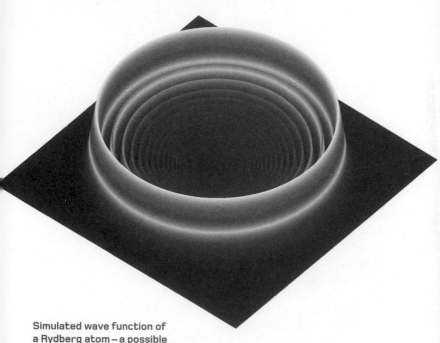

Simulated wave function of
a Rydberg atom — a possible
means of confining photons for
use in quantum computing

Quantum logic gates

A classical computer uses small components called logic gates that carry out simple logical functions based on the electronic signals (bits of binary data) that are fed into them. For example, an AND gate multiplies two inputs, an OR gate adds two inputs and a NOT gate inverts a single input. There are other variations of these gates, but only the NOT gate is reversible; the others are one-way only.

Quantum logic gates, in contrast, are all reversible. Sequences of quantum gates form 'quantum circuits', and since they only perform functions based on one or two inputs, their behaviour can be described in terms of either 2 x 2 or 4 x 4 matrices. There can be many more quantum gates than ordinary logic gates, each performing a different function on the qubits. In the past, quantum gates have been built out of such exotic materials as Rydberg atoms and photons, but in 2015 researchers were able to build a quantum gate out of silicon for the first time, a major step towards making quantum computers practical.

A CNOT gate performs a binary 'NOT' operation on a qubit, flipping its state from 0 to 1 or vice versa, but only if a second control qubit is in state '1'. In 2013 researchers succeeded in building such a gate using a photon and a quantum dot (see page 348).

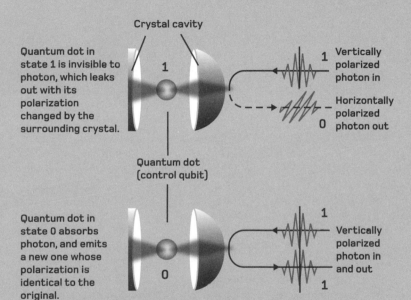

Crystal cavity

Quantum dot in state 1 is invisible to photon, which leaks out with its polarization changed by the surrounding crystal.

1

1 Vertically polarized photon in

Horizontally polarized photon out 0

Quantum dot (control qubit)

Quantum dot in state 0 absorbs photon, and emits a new one whose polarization is identical to the original.

0

1 Vertically polarized photon in and out 1

Quantum algorithms

An algorithm is a step-by-step procedure that tells a
computer how to solve a problem or perform a task.
Normal algorithms can run on quantum computers, but there
are also quantum algorithms specially designed to take
advantage of qubits' inherent ability for parallel processing.

Because these algorithms work on the principle of finding a
solution from one of two answers (0 or 1, true or false and
so on), they can't do anything that is illogical, or theoretically
impossible for a normal computer to accomplish. What they can
do, however, is solve problems much faster. A task that might
take a normal computer centuries might be completed using a
quantum algorithm in a matter of minutes.

The algorithms utilize quantum logic gates to act on a given
number of qubits of input data, culminating in a measurement
that reveals a result. Among the most important are Grover's
algorithm (shown opposite) and Shor's algorithm.

Grover's algorithm

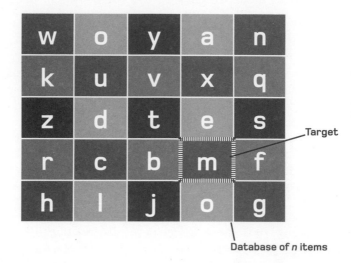

Grover's algorithm is a simple quantum algorithm for sorting through an unordered database and finding a specific item.

Classical solution: n queries required to guarantee finding target.

Quantum solution: √n queries required to guarantee finding target.

Quantum error correction

Quantum computers are so delicate that you can quite literally change the state of qubits by looking at them. Decoherence will inevitably introduce 'noise' and logic gates will introduce occasional errors, just as they do in classical computers. Traditionally, the simplest way to fix such errors has been through redundancy; bits of information are copied and sent multiple times. If there are errors, the repetition may become scrambled and the computer can detect and correct the mistake.

Unfortunately, this procedure doesn't work for qubits. We cannot copy their quantum states because we don't know what they are. This is called the 'no cloning theorem'. However, the information stored on a qubit can be spread across multiple qubits by entanglement, for instance encoding it into the spins of three electrons. Once this is done, errors can be spotted using a so-called 'syndrome measurement' that doesn't disturb the superposition, and suggests various recovery procedures that can fix the errors without causing further decoherence.

Simplified error correction procedure

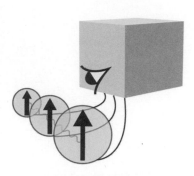

1. Information entangled on three separate qubits

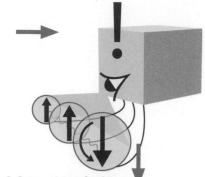

2. Comparison of qubits reveals one has an error

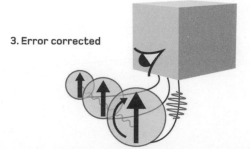

3. Error corrected

Quantum simulations

Unsurprisingly, quantum processes and systems are difficult to simulate in nonquantum computers, so a major application of quantum computing will be to better understand quantum physics itself, along with its applications.

Take, for example, collisions in a particle accelerator like the Large Hadron Collider. A powerful quantum computer could model these collisions in a virtual experiment, showing the energies created and the daughter particles released in high detail before the actual experiment is performed. More exotically, quantum computers could be used to describe conditions at the cores of neutron stars, where the temperatures and pressures are so great that matter could take the form of superconducting superfluids governed by the strong nuclear force. More down-to-earth applications could include allowing greater understanding of high-temperature (that is, closer to 0°C/32°F) superconducting materials, and even designing better quantum computers!

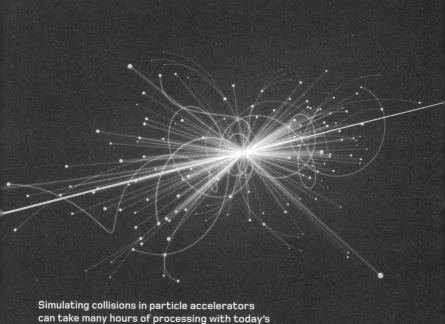

Simulating collisions in particle accelerators
can take many hours of processing with today's
supercomputers, but future quantum machines
may complete similar tasks in minutes.

Building quantum computers

Quantum computing is still in its infancy: the first experimental test of a quantum algorithm took place in 1998 in Oxford, using just two qubits held in an MRI machine. The same year a three-qubit computer was built and, by 2000, the Los Alamos National Laboratory in the United States had a seven-qubit magnetic resonance quantum computer up and running. These systems, however, were so basic that they could only solve the simplest of problems. In 2001, Shor's algorithm was first demonstrated at Stanford University. The quantum computer that achieved this (calculating that the two prime factors of 15 are 3 and 5) also had just seven qubits.

In 2012, however, a Canadian company called D-Wave claimed to have built an 84-qubit computer using quantum annealing, and in 2015 they announced the first 1,000-qubit chip. However, many sceptical scientists have pointed out that, while these may technically be described as quantum computers, in practice they are no faster than ordinary computers.

A 128-qubit quantum computing chip built by Canadian company D-Wave

Future challenges

Quantum mechanics is not some isolated corner of the scientific world. It underlies almost every aspect of physics, chemistry and even biology – from electronics to astrophysics and from medicine to materials science. As we seek to make new advances in these areas and others, our understanding of the quantum realm is sure to play a pivotal role. New technologies, new energy sources and the use of quantum computing will all be part of our quantum future.

But it's not only applications that will develop. Fundamental question marks still remain over the meaning of some crucial aspects of quantum mechanics. Is the wave function a real wave or just an abstract concept? Is human consciousness a key element in defining the quantum reality around us? Can we unify quantum mechanics with other theories? Will we ever learn the origin of the Big Bang and the true nature of the Universe? These are big questions, but if we can crack these quantum challenges, the rewards will be tremendous.

The development of more efficient, cheaper solar cells is just one practical area where the future may be shaped by our growing ability to harness quantum physics.

The observer's role

If the act of observing and measuring a wave function causes it to collapse, this raises a number of philosophical questions. If something is never observed, does it exist, or does it remain in a state of superposition? If the latter is true, then vast expanses of the Universe in which there are no observers would remain as uncollapsed wave functions. Most quantum physicists disregard this avenue of thought, pointing out that the wave function of an object is not the actual physical object itself, but just a way of describing the object's quantum properties. Furthermore, 'measurements' may be made simply by interaction with other particles and radiation, causing decoherence regardless of whether anybody is watching.

But not all physicists are ready to discount the role of the observer. American theoretical physicist John Wheeler argued that the Universe and the observer are mutually dependent; one needs the other, and simply making a measurement is not enough: a conscious mind needs to read that measurement.

Objective collapse

The Copenhagen interpretation is famously noncommittal on some important facets of quantum mechanics. For example, it makes no judgement as to whether the wave function is real, but merely treats it as a description of the probabilistic nature of quantum states. The many-worlds interpretation removes this vagueness by stating that the wave *is* real and branches off into different universes, though this in turn raises many new questions.

These two extremes leave room for other models between them, one of which is objective collapse theory. As its name suggests, this treats the wave function as a real phenomenon, with a collapse that is also objectively real. However, once the wave function has collapsed, that's the end of it; there is no 'branching off' as in many worlds. What's more, objective collapse happens either at random or at a certain scale threshold, with no special role for the observer. However, critics point to a problem that has not yet been resolved; in order for energy to be conserved, a small part of the wave function must somehow remain uncollapsed.

Some scientists argue that objective collapse is a necessity in order for our matter-rich Universe to emerge from the Big Bang.

In the strictest version of the Copenhagen interpretation, the lack of some form of observer in the aftermath of the Big Bang should mean that wave functions do not collapse and matter remains delocalized.

Objective collapse allows wave functions to collapse without an observer. Matter starts to form localized clumps that act as seeds for large-scale cosmic structure.

The early Universe

Recreating the conditions of the Big Bang seem out of reach for any particle accelerator on Earth in the foreseeable future. At present, our best way of understanding the origins of the Universe is to probe the depths of space and observe how astronomical objects were influenced by quantum gravity in the first fraction of a second after the Big Bang. This requires a better understanding of inflation, of dark energy and of the large-scale structure of matter in the Universe today.

The key lies in the cosmic microwave background radiation (CMBR, see page 218), the faint glow of radio waves imprinted with the conditions of the early Universe. The best observations of the CMBR so far came from the European Space Agency's Planck spacecraft (opposite) between 2009 and 2013, but future missions to study the CMBR should be able to rule out up to three-quarters of the potential models for inflation, bringing us closer to understanding the early quantum Universe.

Is information destroyed?

In 1974, Stephen Hawking made his reputation with the discovery that black holes are not quite as inescapable as they may first seem. Virtual particles formed on the edge of a black hole can producing Hawking radiation (see page 242), reducing the black hole's mass until it evaporates completely. None of this Hawking radiation contains information from inside the black hole, however.

When matter falls into a black hole, it contains information in the form of its quantum states, but Hawking (opposite) could see no way for his radiation to conserve this information. He made a bet with US physicist John Preskill that information was destroyed, but conceded the bet in 2004. Why? AdS/CFT correspondence (see page 282) provides a new description of black holes as particles on the boundary between our four dimensions and a fifth dimension, from which our Universe is projected like a hologram. These particles operate by the laws of quantum mechanics, and so *must* conserve information. But exactly how information survives a black hole remains a mystery.

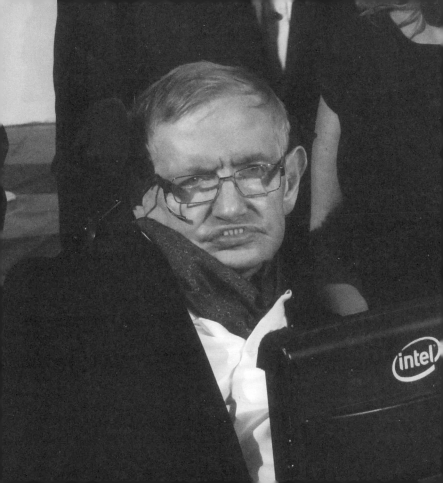

A varying speed of light?

According to conventional physics, the speed of light in a vacuum is a constant 299,792,458 metres per second (186,282 miles per second) wherever you are in the Universe. So why do some scientists suspect it can change? One reason is that, thanks to Heisenberg's uncertainty principle, space is filled with virtual particles. Photons travelling through the vacuum of space will inevitably bump into them, and the energies of the particles could potentially impart a tiny effect on the photons, slowing their speed by a hundred trillionths of a second every metre. Across billions of light years of space, this could build up into a detectable difference.

A varying speed of light has also been suggested as an alternative to the burst of inflation widely believed to have occurred shortly after the Big Bang (see page 214). If fundamental constants can change, then laws of physics can also change with them, presenting all kinds of problems for our understanding of the Universe.

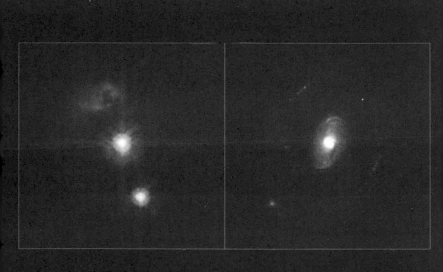

Some astronomers studying light from distant quasars even claim to have found variations in the 'fine-structure constant', a property that defines the strength of electromagnetic interactions between particles and hence also the speed of light.

Extreme matter

Understanding the behaviour of matter under extreme conditions of temperature and pressure remains a huge challenge to physicists. Within the gas-giant planet Jupiter (opposite), pressures 40 million times greater than those on Earth's surface turn hydrogen into an electrically conducting liquid, in which the quantum states of hydrogen atoms could create matter that is both superfluid and superconducting at the same time. Things get even stranger inside neutron stars, where pressures can be 100 billion trillion trillion times greater than on Earth. Some scientists speculate that neutrons could break down into individual quarks, forming a new form of matter that is a plasma of quarks and gluons.

Even greater pressures and temperatures, like those in the Big Bang, are in the realm where different quantum field theories become unified (see page 262). Future particle accelerators, capable of reaching energies of 100 trillion eV (12 times that achievable by the LHC), will aim to probe these conditions.

Alternatives to strings

While string theory (see page 268) is still considered the most likely candidate for a viable theory of everything, not everyone is happy with it. The vast number of potential solutions offered by its equations make it incredibly hard to falsify, since exponents of the theory can suggest any one of 10^{500} potential vacuum states as an alternative to any that are shown to be wrong. Another criticism is that string theory is not 'background dependent': strings vibrate in space and through time, but don't tell us how space and time come to be.

Lee Smolin, the inventor of loop quantum gravity (see page 266), argues that string theory eschews experimental results in favour of elegant mathematics, describing multiple dimensions and parallel worlds without considering whether they can be tested. In 1999, Smolin and backer Mike Lazaridis helped set up the Perimeter Institute for Theoretical Physics in Canada, where researchers are free to investigate not just string theory, but alternative theories of quantum gravity.

	String theory	Loop quantum gravity
Origins	Developed from quantum theory	Developed from general relativity
Scope	Potential theory of everything	Quantum theory of gravity, removing need for theory of everything
Requirements	Supersymmetric particles	No supersymmetric particles
	10 or 11 dimensions	4 dimensions (can incorporate more)
	Continuous spacetime	Quantized spacetime
Possible tests	Graviton particles with spin of 2	White holes (spacetime objects where matter swallowed by black holes escapes)
	Composition-dependent variation in gravity	Varying speed of light at highest energies (gamma rays)

Is Copenhagen right?

The Copenhagen interpretation has dominated quantum mechanics for almost a century, but how literally should we take it? Niels Bohr still considered particles and atoms as inherently deterministic, and viewed the wave function as simply our best attempt to conceptualize it. This works fine for when it comes to solving standard quantum mechanical problems. Explaining Young's interference fringes or the orbitals of electrons doesn't really depend on whether the wave function is a physical thing or an abstract idea; the maths work either way.

But dig a little deeper into nature, and the question becomes more important. The difference between a literal wave and a conceptual wave could be an infinite array of parallel universes or the key to the secret of quantum gravity. The challenge for quantum physicists over the coming decades is to decide which of these concepts is the correct one. Depending on the answer, our understanding not only of nature, but of our place in the Universe (or multiverse), could look very different.

'I am convinced that (God) does not play dice.'

Albert Einstein

Glossary

Alpha particle
A particle released by radioactive decay that consists of two protons and two neutrons – equivalent to the nucleus of a helium atom.

Angular momentum
A property of rotating objects analogous to momentum, and linked to their inertia and rate of rotation around an axis of rotation.

Atom
The smallest indivisible unit of matter that displays the properties of a chemical element. Atoms have a compact nucleus consisting of positively charged protons and uncharged neutrons, surrounded by a cloud of negatively charged electrons whose number balances the number of protons.

Beta particle
A particle released by radioactive beta decay – usually an electron but, rarely, a positron. Beta particles are released from unstable atomic nuclei when a neutron transforms into a proton or, more rarely, vice versa.

Boson
A particle with zero or whole-number 'spin.' Elementary particles known as gauge bosons, often created as virtual particles, play a vital role in transmitting the fundamental forces of nature between fermions.

Electromagnetic radiation
A natural phenomenon consisting of electrical and magnetic waves interfering with and reinforcing one another. It can exhibit very different properties depending

on its wavelength, frequency and energy, and travels in discrete energy packets called photons that display both wave and particle properties.

Electron

A low-mass elementary particle carrying negative electrical charge. Electrons are found in the orbital shells surrounding an atomic nucleus.

Fermion

Any particle with a half-integer spin, including all the elementary matter particles (known as quarks and leptons). Fermions are governed by Pauli's exclusion principle.

Fundamental force

One of four forces governing the way that matter particles interact in nature. Three of the fundamental forces, electromagnetism and the weak and strong forces, are described by quantum physics, but the fourth, gravitation, is currently only described by general relativity.

Gamma radiation

A form of high-energy electromagnetic radiation released by various processes such as radioactive decay.

Heisenberg's uncertainty principle

A relationship stating the impossibility of measuring two 'complementary' quantum properties (for example, a particle's position and momentum) with perfect accuracy at the same time.

Imaginary numbers

A system of numbers based on the square root of -1, denoted *i*. Although *i* does not exist as a 'real' number, it is often required to solve complex equations including many of those describing quantum physics.

Lepton

Any member of a family of elementary particles that are not susceptible to the strong nuclear force, including electrons and neutrinos.

Magnetic moment

A property determining the strength of the magnetic field created by an object, and its susceptibility to the influence of other magnetic fields.

Neutron

An electrically neutral subatomic particle made up of two down quarks and an up quark, found in the nuclei of atoms.

Orbital shell

A region surrounding an atomic nucleus, in which electrons are found. The size of an orbital determines the energy of electrons found there.

Pauli's exclusion principle

A law that prevents fermion particles from occupying identical "states" in a system, and is therefore responsible for much of the structure of matter.

Photon

A discrete packet of electromagnetic energy that can display wavelike, as well as particlelike, behaviour.

Planck's constant

A physical constant that helps define quantum-scale relations such as that between the frequency of a photon and the energy it contains.

Proton

A heavy subatomic particle with positive electric charge, found in the atomic nucleus and composed of two up quarks and a down quark.

Quantum

The minimum possible amount of a particular property that may be involved in a physical interaction. Certain phenomena, such as the energies of light waves and of electrons in an atom, are inherently 'quantized' on the smallest scale. Quantum physics describes the strange and sometimes counterintuitive behaviour that arises as a result.

Quark

An elementary particle found in six different 'flavours', responsible for most of the mass in matter.

Spectral lines
Lines in a spectrum of light with specific wavelengths, caused by the emission or absorption of light as electrons move between orbital shells and energy levels within atoms.

Spin
A property of subatomic particles, analogous to angular momentum in larger objects, which affects many aspects of their behaviour.

Wave function
A description of the quantum state of a system, often denoted by the Greek letter ψ (psi). The wave function describes the probability of a measurement performed on a quantum system producing a particular result.

Vector
A mathematical object with both a magnitude and a specified direction. Many quantum properties are described in vector terms.

Virtual particle
A particle that spontaneously comes into existence and exists for an extremely short time as a result of Heisenberg's uncertainty principle as it applies to time and energy. Virtual particles are produced as particle-antiparticle pairs, and act as gauge bosons transmitting the fundamental forces of nature.

Scientific notation:
This book inevitably deals with some very large and very small numbers. To simplify their presentation, scientific notation is used where appropriate, with numbers presented in the form $a \times 10^b$ (that is, a times 10 to the power of b). Hence $3 \times 10^6 = 3,000,000$ (3 followed by six zeroes). In this system, negative values of b indicate multiplication by $1/10^b$, so for example $3 \times 10^{-6} = 3 \times 0.000001 = 0.000003$.

Index

AdS/CFT correspondence 282, 398
aether 10, 16
algorithms, quantum 382, 388
alpha decay 128, 130, 168
anthropic principle 302
antimatter 164, 166
applications, quantum 326—68
atomic clocks 340
atomic structure 42, 44, 46, 58—60

baryons 110, 124
Bell's theorem 308
beta decay 126, 128, 132
Big Bang 214, 216, 224, 230, 250, 280, 300, 396, 402
'Big Crunch' 248, 250, 300
biological compasses 358
biology, quantum 356
black body radiation 24, 26, 28, 34
black holes 240, 242, 398
Bohr, Niels 44, 54, 178, 182, 206, 406

Boltzmann brains 324
Born, Max 188, 150
Born rule 150
Bose-Einstein condensates 112, 114, 350
bosons 112, 114, 120, 126, 136, 274, 350
brain 364, 366
brane theory 280, 300

Calabi-Yau space 278
carbon dating 346
Casimir effect 136
cathode rays 30
chemistry, quantum 354
chirality 104, 126
clocks, quantum 340
complementarity 182
Compton scattering 36
computing, quantum 370—88
conformal field theories (CFTs) 282
consciousness, quantum 364, 366
Copenhagen interpretation 54, 144,

146, 156, 178, 180, 286, 292, 304, 306, 394, 406
corpuscular theory 12, 14
correspondence principle 206, 208
cosmic microwave background radiation (CMBR) 218, 396
cosmological constant 232, 282
Coulomb barrier 130, 168, 170
cryptography, quantum 342, 344
cyclical universes 300

dark energy 230, 232, 248, 282
dark matter 98
de Broglie wavelength 38, 40, 46
decoherence, quantum 176, 376, 378, 384
degeneracy, quantum 72, 236, 238
determinism 312, 368
dice, playing 304
Dirac equation 164

Dirac, Paul 164, 194, 200
Doppler effect 228
dots, quantum 348, 378
double-slit experiment 12, 14, 154, 200, 208, 322

$E=mc^2$ equation 50, 160, 216
eigenfunctions 198, 204
Einstein, Albert 28, 32, 34, 48, 50, 52, 114, 146, 160, 304, 306, 310, 314
electric charge 100
electromagnetic force 18, 20, 122, 254, 258, 264, 272
electron tunnelling microscope 330
electronics 334
electron(s) 30, 34, 44, 46, 58, 60—80, 100, 104, 106, 108, 174
    diffraction 40
    energy levels 64—6
    excited 68, 74, 134
    forbidden transitions 83
    shells 60, 62, 68, 70, 76
    subshells 62, 70
electroweak theory 264
emission lines 80
energy, conservation of 82
entanglement, quantum 306, 308, 310, 312, 314, 316, 320, 376

entropy 22, 320, 324
EPR paradox 306, 308
error correction, quantum 384
Everett, Hugh 286, 292
expanding Universe 214, 228, 230, 234
extreme matter 402
eyes 362

Faraday, Michael 18, 106
fermions 110, 274
Feynman diagrams 202, 258
Feynman, Richard 200
flash drives 336
Fraunhofer lines 78
free will 368
future of quantum physics 390—406

galaxies, origins of 220
gamma decay 134
general relativity 48, 206, 240, 252, 262
gluons 94, 112, 136
grand unified theory (GUT) 244, 270
graviton 262
gravity, quantum 240, 248, 250, 252, 262, 266, 272

ground-state electrons 68, 74
Guth, Alan 224, 226

hadrons 94, 116
Hamiltonian operator 196, 198, 204, 210
harmonic oscillators, quantum 158, 210
Hawking radiation 242, 398
Heisenberg, Werner 46, 52, 172, 188
Heisenberg's uncertainty principle 52, 140, 172, 174, 182, 242, 254, 262, 304, 312, 320, 324, 400
hidden variables 308
Higgs boson 88, 118, 120
higher dimensions 276
Hilbert space 192, 194
horizon problem 222, 224
Hubble, Edwin 228
Hund's rules 74, 76
Huygens, Christiaan 10, 16

inflation 224, 226, 246, 250, 290, 300, 400
inflationary multiverse 290

Klein-Gordon equation 162, 164

Lamb shift 138

Large Hadron Collider (LHC) 116, 118, 166, 284, 386
lasers 328, 344
leptons 96, 110
light emitting diodes (LEDs) 338
light, speed of 314, 400
limits of the quantum realm 208
logic gates, quantum 380, 382, 384
loop quantum gravity (LQG) 266, 284, 404

magnetic moment 102, 106, 108
magnetic resonance imaging (MRI) 174, 332
many-worlds interpretation 54, 144, 286, 292, 294, 296, 298, 316, 394
mass-energy equivalence 48, 50
mathematics, quantum 184—210
matrix/matrix mechanics 186, 188, 190, 194
Maxwell, James Clerk 18
Maxwell's equations 20
memory stick 336
mesons 112, 124

Michelson-Morley experiment 16
momentum, conservation of 82
Mozi satellite 344
multiverses 286—302

neutrinos 96, 98, 100
neutron stars 234, 236, 238, 386, 402
neutrons 42, 58, 100, 126, 132
Newton, Isaac 10, 12
nucleus, atomic 58, 60

objective collapse theory 394
observable Universe 288
observer's role 392
operators, quantum 196, 198, 204

parity 104
particle angular momentum 102
particle physics 88—140
particle theory 12, 14, 36, 38
path integral formulation 200
Pauli's exclusion principle 76, 94, 110, 236, 238, 260, 348

Penrose, Roger 364
perturbation theory 210
photoelectric effect 32, 34
photons 32, 34, 36, 122, 136, 138, 174, 258, 362, 400
photosynthesis, quantum 360
Planck epoch 270, 272, 274
Planck, Max 24, 26, 28, 32, 34
Planck relation 44
Planck constant 28, 38, 66
positrons 100, 166
probabilities, quantum 148, 150
probability wave function 144
protons 42, 58, 100, 132, 332
  decay of 244

quanta 28
quantum chromodynamics (QCD) 94, 260
quantum electrodynamics (QED) 258
quantum field theory (QFT) 254, 402
quantum fluctuations 216
quantum mechanical atom 46
quantum numbers 62, 152

quantum states 152, 156
quantum suicide 294, 296
quark stars 238
quarks 58, 90, 92, 94, 110, 124, 132, 260
qubits 370, 372, 376, 378, 384
quintessence 232

radioactivity 128, 130, 132, 134, 346
radiometric dating 346
relativity 48, 160, 162, 206, 214, 240, 252, 262, 314
Rutherford, Ernest 42, 44
Rydberg constant 66

Schrödinger's cat 178, 180, 294, 296
Schrödinger's wave equation 70, 156, 162, 188, 190, 194, 198, 204, 210, 254
simulations, quantum 386
Solvay Conference (1927) 52
special relativity 48, 160, 162, 314
spectroscopy 44, 56
spin, quantum 62, 86, 102, 104, 106, 110, 332
spin-orbit interactions 108

Standard Model 88, 90
Stark effect 210
stars 24, 234—8
    death of 234
string theory 192, 268, 274, 276, 278, 284, 290, 404
strong force 124, 260, 272
superconductivity 352
superfluids 350
supernovae 230, 234
superpositions, quantum 154, 176, 178
superstrings 106, 278
supersymmetry 118, 274, 276, 284
symmetry 256
symmetry breaking 272

Tegmark, Max 288, 290
telecommunications 344
teleportation, quantum 316—18
'theory of everything' 192, 252—84, 404
thermodynamics 22, 82
Thomson, J.J. 30, 42
time, quantum 320—4
time running backwards 322
transformation theory 194

tunnelling, quantum 130, 148, 168, 170, 174, 330, 356
ultraviolet catastrophe 26, 32
uncollapsible wave function 286, 292, 296, 392, 394
Universe 212—50, 396
    fates of the 248

vacuum decay 246
vacuum energy 140, 232
virtual particles 136, 140, 174, 214, 216, 242, 254, 398, 400
vision, quantum 362

wave function 142—82, 190, 200, 204, 286, 292, 392, 394, 406
wave-particle duality 38, 40, 42, 46, 142, 188, 208
wave theory 10, 12, 14, 16, 18, 20, 38
weak force 126, 264, 272

Young, Thomas 14, 16, 146

Zeeman effect 62, 86

Quercus

New York · London

Text © 2017 by Gemma Lavender
First published in the United States by
Quercus in 2017

ISBN 978-1-68144-174-0

Library of Congress Control Number:
2017931428

Distributed in the United States and Canada
by Hachette Book Group
1290 Avenue of the Americas
New York, NY 10104

Manufactured in China

10 9 8 7 6 5 4 3 2 1

www.quercus.com

Picture credits 2: general-fmv via Shutterstock, Inc.; 11: Fouad A. Saad
via Shutterstock, Inc.; 13: Mopic via Shutterstock, Inc.; 19: Fouad A. Saad
via Shutterstock, Inc.; 71: Patricia.fidi via Wikimedia; 79: N.A.Sharp, NOAO/
NSO/Kitt Peak FTS/AURA/NSF; 81: ESO/H. Drass et al.; 85: NASA; 89:
CERN/BEBC; 99: ESA/Hubble & NASA; 109: Johnwalton via Wikimedia;
115: Courtesy National Institute of Standards and Technology; 119: Julian
Herzog via Wikimedia; 121: CERN for the ATLAS and CMS Collaborations;
123: FurryScaly/ Flickr; 143: BenBritton via Wikimedia; 153: PoorLeno via
Wikimedia; 159: Alexander Sakhatovsky via Shutterstock, Inc.; 161: NASA;
181: Rhoeo via Shutterstock, Inc.; 189: GFHund via Wikimedia; 193: Jorge
Stolfi via Wikimedia; 201: Matt McIrvin via Wikimedia; 211: Michael Courtney
via Wikimedia; 213: NASA, ESA, S. Beckwith (STScI) and the HUDF Team; 217:
GiroScience via Shutterstock, Inc.; 219: NASA / WMAP Science Team; 221:
Max Planck Institute for Astrophysics/Springel et al., 2005; 227: Juergen
Faelchle via Shutterstock, Inc.; 237: NASA/CXC/SAO/F.Seward et al.; 247:
Vadim Sadovski via Shutterstock, Inc.; 251: Mikkel Juul Jensen/Science
Photo Library; 255: Tenth Seal via Shutterstock, Inc.; 263: Markus Gann
via Shutterstock, Inc.; 265: CMS/CERN; 269: DrHitch via Shutterstock,
Inc.; 273: r.classen via Shutterstock, Inc.; 279: Lunch via Wikimedia; 283:
Fritz Goro/Contributor/Getty Images; 287: Detlev van Ravensway/
Science Photo Library; 291: Henning Dalhoff/Science Photo Library; 295:
Rhoeo via Shutterstock, Inc.; 297: ktsdesign via Shutterstock, Inc.; 299:
Victor Habbick via Shutterstock, Inc.; 301: Nicolle R. Fuller/Science Photo
Library; 305: Mega Pixel via Shutterstock, Inc.; 313: PHOTOCREO Michal
Bednarek via Shutterstock, Inc.; 319: Volker Steger/Science Photo Library;
321: Svetlana Lukienko via Shutterstock, Inc.; 325: Johan Swanepoel via
Shutterstock, Inc.; 327: ORNL/Jill Hemman; 329: Designua via Shutterstock,
Inc.; 335: Robert Lucian Crusitu via Shutterstock, Inc.; 341: ESA - J. Huart;
345: Sebastian Kaulitzki via Shutterstock, Inc.; 355: Simen Reine-UIO;
357: Andrii Vodolazhskyi via Shutterstock, Inc.; 359: FotoRequest via
Shutterstock, Inc.; 361: Smit via Shutterstock, Inc.; 363: Roland Deschain
via Wikimedia; 365: Alila Medical Media via Shutterstock, Inc.; 367: AkeSak
via Shutterstock, Inc.; 371: Mopic via Shutterstock, Inc.; 379: Hweimer
via Wikimedia; 387: general-fmv via Shutterstock, Inc.; 389: D-Wave
Systems, Inc. via Wikimedia; 391: Fomanu via Wikimedia; 393: Aleksandar
Mijatovic via Shutterstock, Inc.; 397: ESA - AOES Medialab; 399: Martin
Hoscik via Shutterstock, Inc.; 401: NASA/ESA, ESO, Frédéric Courbin
(Ecole Polytechnique Federale de Lausanne, Switzerland) & Pierre Magain
(Universite de Liege, Belgium); 403: NASA/JPL/Space Science Institute.
All other illustrations by Tim Brown.